The Evolution of Imperfection

The Evolution
of Imperfection

THE SCIENCE OF WHY WE AREN'T
AND CAN'T BE PERFECT

LAURENCE D. HURST

PRINCETON UNIVERSITY PRESS

PRINCETON & OXFORD

Requests for permission to reproduce material from this work should be sent to permissions@press.princeton.edu

Published by Princeton University Press
41 William Street, Princeton, New Jersey 08540
99 Banbury Road, Oxford OX2 6JX

press.princeton.edu

All Rights Reserved

ISBN 9780691247397
ISBN (e-book) 9780691247410

British Library Cataloging-in-Publication Data is available

Editorial: Alison Kalett and Hallie Schaeffer
Production Editorial: Theresa Liu
Jacket/Cover Design: Karl Spurzem
Production: Erin Suydam
Publicity: Matthew Taylor and Kate Farquhar-Thomson
Copyeditor: Annie Gottlieb

This book has been composed in Arno and Sans

Printed in the United States of America

10 9 8 7 6 5 4 3 2 1

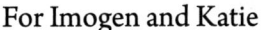

For Imogen and Katie

The man with insight enough to admit his limitations comes nearest to perfection.

—JOHANN WOLFGANG VON GOETHE

CONTENTS

PREFACE

This book is the product of a long-term fascination, born first from inspirational lecturers (Nick Davies and Tim Clutton-Brock) and then from mentors who were variously visionary (Bill Hamilton) and clear-thinking (Alan Grafen). From these I developed a respect for—and love of—a good question well posed. From Bill I especially developed a love of scholarship—reading the literature as a means to develop and test ideas. Thanks must also go to Jonathan Milner for encouraging me to venture into science communication.

More immediately, this book is the product of remarkable good fortune. In December 2020, after a hard term of online COVID teaching, I was asked by the Wissenschaftskolleg (Institute for Advanced Study) zu Berlin—familiarly known as the Wiko—to appraise the case of a potential visiting fellow. This I was delighted to do, and I added a cheeky "P.S. Can I also come and visit?" expecting nothing of it. Between Permanent Fellow Dieter Ebert and the Rector, Barbara Stollberg-Rilinger, somehow a very late invitation for a short stay was made possible. This offer came as a great relief, such was the difficulty of teaching and researching through COVID. To both Dieter and Barbara, a most heartfelt thanks. I suspect the disruption owing to COVID was part of the reason openings were available. If so, I have SARS-CoV-2 to thank for the opportunity.

And yet, because of the virus, it very nearly never happened. Germany not only shut its borders the summer before my visit (2021), but owing to the spread in the UK of a novel variant (Delta, if I recall), UK residents were not even considered for a visa. Fortunately, just in time, the German government relaxed their restrictions (thanks to them), and with great assistance from the Wiko staff (thanks especially to Vera

Pfeffer), a visa was obtained. When initially accepting the offer of a short-term stay, I had no idea what I would do. But at Wiko's request I wrote a paragraph on my interests, confidently expecting no one to read it. Fortunately, Alison Kalett saw it and contacted me out of the blue suggesting it might make a good book. And so, by further lucky happenstance, shortly before I took up residence, I had a good reason to be in Berlin. As I have never written a book before, Alison's guiding hand has been invaluable.

This book was developed, and in part written, during those fortunate months. I have my Wiko colleagues to thank for stimulating discussion. Special thanks go to Rachel Wheatley, who seemed to know when I was going slightly deranged, accompanying me to pick up my morning croissant. Two anonymous referees provided both great encouragement and most helpful critiques, for which thanks. The greatest thanks, however, must go to my long-suffering wife, Clair, who took over all dog-walking duties while I remained indulgently hermit-like in Berlin.

ABBREVIATIONS

A   adenine

AAV   adeno-associated virus

AIDS   acquired immunodeficiency syndrome

ADR   adverse drug reaction

BGI   Beijing Genomics Institute

bp   base pair

C   cytosine

CDS   coding sequence

CRISPR   clustered regularly interspaced short palindromic repeats

CRP   C-reactive protein

C-SECTION   Cesarean section

CVS   chorionic villus sampling

DMD   Duchenne muscular dystrophy

DNA   deoxyribonucleic acid

DSM   *Diagnostic and Statistical Manual of Mental Disorders*

ENCODE   Encyclopedia of DNA Elements

ESE   exonic splicing enhancer

EU   European Union

FDA   Food and Drug Administration

G   guanine

gBGC   GC-biased gene conversion

GRS genetic risk score

GWAS genome-wide association study

HIV human immunodeficiency virus(es)

IVF in vitro fertilization

Ka number of nonsynonymous changes per nonsynonymous site

kb kilobase

Ks number of synonymous changes per synonymous site

LDL-C low-density lipoprotein version of cholesterol

LTEE long-term evolution experiment

Mb megabase

mg milligram

MR Mendelian randomization

mRNA messenger RNA

μ (mu) mutation rate

NCBI National Center for Biotechnology Information

N_e effective population size

NHS National Health Service

NIH National Institutes of Health

NMD nonsense-mediated decay

PGD pre-implantation genetic diagnosis

PRS polygenic risk score

PSA prostate-specific antigen

RCT randomized controlled trial

RIP repeat induced point mutation

RNA ribonucleic acid

RONS reactive oxygen and nitrogen species

s selection coefficient

SJS   Stevens-Johnson syndrome

SRG   Statistical Research Group

T   thymine

TEN   toxic epidermal necrolysis

UK   United Kingdom (of Great Britain and Northern Ireland)

UPD   uniparental disomy

US   United States (of America)

WHO   World Health Organization

The Evolution of Imperfection

1

Introduction:
The Problem of Perfection

Evolution.

Reading that, what came to your mind? Asking this question of many British fifteen-year-olds, the most common responses were "Darwin," "monkey," "adapt," and "natural selection." Being that it was the largely secular UK, not many mentioned anything to do with religion, but I'm guessing that might be different elsewhere. So, what came into your head? Was it one of those answers? Or perhaps you thought of "revolution"? "Evolution not revolution" is one of those metaphors beloved of technology journalists.

And what image comes to mind?

You would be unusual if you didn't think of Rudolph Zallinger's classic portrayal of evolution: starting with a chimp-like ape, crouched over, knuckle walking, and then, viewed left to right, gradually becoming more human: standing up, getting taller, and becoming more hairless. Put the word "evolution" into Google image search and most of the top hits are this image or a variant of it.

What many of these first responses have in common is the idea that evolution is a process of gradual improvement, with us (humans) as the pinnacle of evolution. Indeed, Zallinger's image, originally titled "The Road to *Homo sapiens*," is more commonly referred to as "The March of Progress." In the image, the various forms are all striding left to right: the direction of travel, literally and metaphorically, is clear. The increase in height from left to right visually reinforces that same idea of our superiority and of progress.

Is evolution simply a process of gradual improvement, a progressive march toward perfection rendering us the finest nature has to offer? Shakespeare's Hamlet thought so, declaring: "*What a piece of work is a man, how noble in reason, how infinite in faculties, in form and moving how express and admirable; in action how like an angel, in apprehension how like a god: the beauty of the world, the paragon of animals . . .*"

There are reasons to think that Hamlet and Zallinger may be onto something. The process of natural selection, survival of the fittest, envisages one type that better fits the environment outcompeting some less well-adapted type. Repeating such a process over and over should lead to a species fitting its environment like a hand in a custom-made glove. All living species should then be better than their recent ancestors, who should in turn be an improvement on their ancestors.

Indeed, when I first heard about evolution by natural selection I was, to say the least, underwhelmed, as the process—and the notion of progress implicit in it—seemed so obvious. I was shown pictures of black moths on dark tree bark, turned sooty by industrial pollution, and a white version of the same moth species on the same dark background. Unsurprisingly, the white type stood out like a sore thumb. And then I was told that, because of the difference in visibility, black moths on a black background were less likely to be eaten by birds than the white version on the same background. Really! Who would have guessed? And then I was told that, because of not being eaten as often, over many generations after the Industrial Revolution the black version became more common than the white version. You don't say.

This evidence that evolution can occur by natural selection, I was also told, was apparently key to our understanding of the world around us and, in particular, to understanding why organisms are so exquisitely adapted to their environment. Generation after generation, those best fitting the environment survive, and so, over time, nature gradually continues to improve until no more improvement is possible: perfection. "Evolution not revolution" captures this steady march toward perfection rather well. So obvious is this that, at first sight, it seems that the idea of progress and perfection is somehow hard-baked into the process of evolution.

The Perfection of Nature

When evolutionary biologists talk about evolution, we also often emphasize this perfecting nature of the process. I still find the remarkable feats of camouflage breathtaking. The appropriately named leaf-tailed gecko looks like a dead leaf in both body and tail. When curled up, its mottled brown body disguises itself wonderfully among twisted brown dry leaves. Similarly, I think you will never spot the tulip tree beauty moth when it blends into the tree bark it rests on. And you might have heard a screech owl (hence the name), but you would have a hard time spotting one poking its head out of its tree hole, their mixed white and black feathers match the mottled bark so well. Indeed, I wonder how many species are still unknown because we cannot see them. The pink-and-white pygmy seahorse is so well disguised against coral that it was only discovered when it hitched a ride to a scientist's laboratory on a coral sample.

Camouflage makes for a visually arresting demonstration of the power of natural selection, but many other examples are differently hidden from view. One of my favorite species is a fungus that digests trees from the inside and then breaks through the trunk when it is ready, forming a sort of half-moon-shaped bracket on the side of the tree. This bracket is there to make and release spores of the fungus, so continuing the cycle: digest tree, find new tree, digest tree, etc. Look underneath this bracket and you see a myriad of fine pores on the underneath, these being the ends of very narrow (about 1 mm) but relatively very long (about 10 cm) tubes (hence the name of this sort of fungus, a bracket polypore). The reason for these long narrow tubes is to maintain a water-rich microenvironment at the top of the tubes.

At the point in the fungus life cycle when they release spores, fungi need a moist environment. They flip the spores out using water pressure—a bit like a water pistol. Unlike their evolutionary relatives, edible field mushrooms, which usually only appear aboveground to release their spores after it has rained, bracket polypores are present year-round, and so have an especially big problem keeping moisture trapped inside during the hot summer months. The solution is for the spores to be released at the top of the tubes under the waterproof hard woody upper layer. Because of the long thin tubes, the moisture where the spores are released can stay high, even on a hot day, in the fungus's own damp microclimate.

While this solves the moisture problem, the length and slenderness of the tubes produce a problem of their own: how to get a spore, which often needs to be sticky to adhere to the next tree, down a long thin tube without touching the sides. Any kink in the tube, and the tube will just clog up and spores will not get released to the open air. The spore needs to be released in just the right way, and the tube needs to be vertical.

And this is where we find a remarkable hidden feat of narrow tube construction. For one of the larger bracket polypore species, *Ganoderma applanatum*, it has been estimated that, for each tube, this feat of engineering is equivalent to building a household drainpipe the height of the Eiffel Tower that is so perfectly vertical that a ball bearing can be dropped down it without touching the sides once. If you ever feel a bit sadistic, take a rotting log with one of these fungi growing out of it and turn it. This is cruelty to fungi, as the tubes are no longer perfectly vertical. But come back a few months later and you will find that the fungus has adjusted and now all its tubes are once again perfectly vertical.

I could go on endlessly about the amazing perfection of so much of nature. Did you know that birds of prey have two lenses in their eyes so they can see a tiny mouse in detail while hovering many meters above the ground? It is like they have a built-in telescope (or binoculars, as both eyes are like this). And did you know that the bee orchid not only has a flower that looks like the back of a female bee but also releases chemicals that mimic the scent of the female bee? All this is to tempt a male bee to come and "mate" with the mimic female bee (i.e., the flower) and so distribute the pollen. The process is known as pseudocopulation, meaning false mating.

The Problem of Imperfection

Given the commonness of apparent perfection in nature, it might seem somewhat perverse to be writing a book about the evolution of imperfection. Isn't it like having the best meal ever cooked for you and complaining that the napkins weren't folded to your liking? In no small part, however, the intellectual problem of imperfection exists because the process of natural selection seems to so inevitably lead to perfection that the absence of perfection becomes an intriguing quandary. Put differently, if the process of evolution is so simple, with repeated bouts of the survival of the fittest being the only important mechanism, how come so many things seem less than perfect? Here I speak as a sixty-something male with a bad back, dodgy knees, failing eyesight, receding hairline, and frankly not much to look forward to. Not exactly a prime example of evolutionary perfection, you might say. Why have we evolved to age like this—and why are so many plagued by a bad back—when, for example, the giant redwood in my garden seems to improve with every passing year?

Sometimes the problem of imperfection is used to attack the idea of evolution. The conversation usually starts, *"If evolution is true, how come..."* You can fill in the blanks here, but it includes everything from men having nipples to monkeys still being around when, implicitly, they are less perfect than we are. But you can also turn it into an intriguing problem: How, if evolution works by survival of the fittest, can it lead to imperfection?

The field of evolutionary biology has provided a diverse series of explanations for apparent imperfections. In some cases, the seemingly imperfect is the best we can do within our constraints (this seems to explain why I age but my giant redwood goes on). In other cases, we suspect there to be a lag between the environment changing and organisms adapting (which may well explain my bad back). Similarly, the plight of well-adapted organisms finding themselves suddenly in the wrong environment seems to explain rising incidences of many conditions such as diabetes and allergies. In other cases, where you start from or how you are genetically wired limits where you can go (which may explain why

I as a male am needed for reproductive purposes). I'll consider these various explanations in more detail later. I'll also argue that for the most part these classical explanations are insufficient in light of new discoveries. But first, what, exactly, might we mean by imperfection?

Imperfection: A Slippery Concept

At its core is the notion that there is an alternative that is better. But what do we mean by "better," and what is an alternative?

"Better" can be a politically and emotionally charged term because the notion of a worse state can be insulting (or worse, as we shall see). Implicit in the concept of a genetic "disease" is the idea of imperfection and that there is a disease-free better state—why, after all, would we try to treat something if it isn't imperfect? It might seem obvious that something like a childhood cancer is a disease that we should cure if we can. However, what is a disorder and what is not is not always so well-defined: we often disagree on whether, when comparing two states, one is better than the other, or whether they are just different. Autism is a case in point. On the one hand, the National Health Service (NHS) in the UK takes the stance that autism is not an illness. The charity Child Autism UK agrees, considering that it should rather be regarded as a difference in information and stimulus processing, not a disease and not a condition requiring a cure. Despite this, "autism spectrum disorder" is a medical diagnosis, criteria for which are set out in the *Diagnostic and Statistical Manual of Mental Disorders* (DSM-5) from the American Psychiatric Association. Note the repeated use of the word "disorder." Similarly, the Mayo Clinic talks of reducing "*symptoms*." Were I a parent of a possibly affected kid, I would be confused: How can it be both "not an illness" and also a disorder with symptoms? The fact that so many quacks have set up businesses promising "cures" for autism would suggest that many parents consider it anything but a form of neurodiversity. How desperate must a parent be to give their child the purported cure "Mineral Miracle Solution," which turns out to be bleach? It doesn't follow that all standpoints are equally valid, but it does mean that "imperfection" may be a loaded term.

Fortunately, for our purposes we can, at least in principle, try to short-circuit this problem. We can evoke the idea that *from an evolutionary point of view* some traits are selected against and so must be evolutionarily imperfect: the white moth would appear to be imperfect if the trees are all black.

The other difficulty is the problem of the "alternative" possibilities. Do we mean that if I can *imagine* something that is better, then what we have is somehow imperfect? I could imagine great advantages to being able to run faster than any predator that might eat me. Alternatively, perhaps I mean that another species seems to have arrived at a better solution. As we shall see, our eye has a strange quirk—literally a blind spot—not seen in the octopus eye. Is ours therefore imperfect? Perhaps I mean something is perfect given my constraints. If I give you some small amount of money to buy a car, you could well come back with the best you could do for the price, just not the best car. Is your car perfect? Imperfection seems rather hard to define.

As imperfection is a bit of a slippery concept, why, you might ask, would anyone study something that possibly defies clean definition? If that is how you feel, please don't put the book down just yet.

My premise for this book is simple: it is by studying the cases where what is seen doesn't obviously make sense, and thus appears less than perfect, that we might come to a fuller understanding of the evolutionary process. The study of imperfection has more to do with finding interesting questions than with making definitive statements about whether something is or is not perfect.

I am interested, then, in imperfection in the sense that some things do not make *obvious* sense when we start from the presumption that evolution is a process enabling constant improvement until an endpoint of perfection. I, for example, don't find black moths surviving on black backgrounds very interesting, as it is too obvious. I am similarly not very interested in why the lens of the eye is transparent—as all sufferers of cataracts will tell you, it would be a lousy eye if this wasn't the case. I am, conversely, interested in cases that don't look so obvious, such as why we need so many sperm to fertilize one egg, why most human fertilized eggs never make it, and why much of our DNA appears to be rather pointless. How the answers in turn change our view of evolution, and how this new view enriches—at least for me—what it means to be human, I hope to convey in this book.

I am not alone in this pursuit: much evolutionary research gravitates toward the same sorts of issues for similar reasons. Sometimes they illuminate problems that you might not have thought were problems. Why organisms reproduce sexually rather than asexually (where females make only daughters without a contribution from males), and why individuals can sometimes be kind to others at a cost to themselves (altruism), have historically been two of the central problems of my field. From an evolutionary point of view they appear to be head-scratching imperfections. In the case of sexual reproduction, an asexual mother could have twice as many daughters as a sexual female, daughters who in turn would have twice as many again. The sexual female is, as we shall see, in the bind of having to make sons. As males in many sexual species don't contribute resources to the kids (just their genes), the asexual lineage can expand from 1 to 2 to 4 to 8 to 16, etc., while the sexual lineages remain the same size. Sexual reproduction should, according to the above logic, be displaced by asexual lineages very rapidly. Selection should favor asexuality. But asexuality (at least obligate asexuality) seems to be rare in multicellular species, of which we are one. Altruism is equally perplexing, as at first sight, in a competitive world, costly giving seems odd. It would be as odd to an economist to see people work hard and then give all their earnings to strangers.

Imperfection Is Especially in
Evidence at the Genetic Level

This book, then, is about those features of organisms that demand an explanation, as they appear not to make sense. The same could have been said, however, anytime in the last 150 years of evolutionary research. What is new—and the focus of this book—is that we currently face a barrage of novel genetic problems, as a result of the new data now pouring out at unbelievable rates about DNA. While at the level of what things *look like* we may see abundant glorious perfection (or something so close that it would be churlish to quibble), it is far from clear that the same perfection is seen at the level of DNA.

Since the human genome (our complete set of DNA) was first sequenced in 2003, we have learned much about the oddness of DNA in both its anatomy and its behavior. We have many fewer protein-coding genes than expected, and, more generally, remarkably little of our DNA (only about 1.2%) does the canonical job of coding for proteins, the "doing" molecules in our cells. However, most of our DNA is active; it just isn't active in the way we expected.

On top of this, for every baby born, about two never made it. In most cases the mother never even knew she was pregnant. Those offspring that do make it to term have on average more changes to their DNA— changes that are more likely to be harmful than beneficial—than are seen in just about any other species.

These and numerous other genetic features all appear at first sight to be imperfections. We seem, therefore, to be missing something about the evolutionary process. Just as our ability to examine DNA and genetics has uncovered the new issues, so too the same technology provides new data to enable us to better understand what is going on. What follows in this book is my attempt to synthesize why we—and mammals more generally—appear to be so very genetically imperfect. It just so happens that considering our genetics reveals processes of evolution that go beyond the simple narrative of the March of Progress.

Evolutionary Imperfection Is Not an Ethical Statement

Before we go there, let's tackle the obvious and difficult question: In suggesting that some genetic features might be imperfect, am I also questioning the moral worth of some people? Once people thought so. Eugenicists of the early twentieth century made the presumed genetic inferiority (alias "impurity") of some people their justification for sterilizing or killing them. Eugenicists would usually suggest that they were somehow purging the gene pool, thus evolutionarily improving humans. Often such policies went hand in hand with immigration (or forced emigration) policies, all on the pretext of keeping gene pools unsullied.

Eugenics wasn't just practiced in Nazi Germany. Its forerunner happened in the United States in plain sight. In deciding the case of Buck v. Bell in 1927, the US Supreme Court, no less, upheld a state's right to forcibly sterilize a person deemed unfit to have kids. The decision, at 8 to 1, wasn't even close. In this case the state had deemed the individual, Carrie Buck, to be "mentally deficient." The deaf and the blind were similarly targeted. The poor, minorities, and women considered to be "promiscuous" were often the victims. At the very least, many tens of thousands of US citizens were forcibly sterilized in the twentieth century. In the US the last case of forced sterilization was in (checks notes) 2021.

Despite a postwar recognition of the horrors of eugenics, implied genetic and evolutionary demonization persists like hardened gum stuck to the sole of your shoe, no matter what counterarguments are presented. There are, for example, common views that conflate judgments of what is "natural" or "unnatural" with what is ethically right or wrong. One only need look at the stigmatization of homosexuality to see such arguments in play. Often these positions start from the assertion that homosexuality must be an imperfection, evolutionarily speaking, as same-sex partners cannot have kids. It is usually then argued that it is unnatural, which is then equated with ethical inferiority, a deficiency in a pejorative sense. The consequences, in the UK at least, were criminalization or forced chemical castration (or both). One of the more famous victims of this system was the great mathematician and code breaker Alan Turing, who ultimately and tragically committed suicide in 1954. In the UK homosexuality was legalized in 1967, and Turing was posthumously pardoned in 2013. In 1973 the American Psychiatric Association's great handbook, the *Diagnostic and Statistical Manual of Mental Disorders* (then DSM-III), removed homosexuality as a mental disorder. In many places it remains stigmatized, illegal, or medicalized.

In the case of homosexuality, we can criticize this logic (and assumed facts) at every step, but that doesn't address the more general point. In this book, I wish to examine questions such as why we have a high mutation rate causing genetic diseases and why so many humans as embryos have an extra chromosome (as in, e.g., Down syndrome). In even considering these as imperfections, am I implicitly endorsing a sort of eugenics or stigmatization?

The answer is a profound no.

There are just so many ways these lines of reasoning are wrongheaded. As moral philosophers have noted, usually the ethical position (or prejudice) comes first and the "defense" for it comes along as an afterthought. Consequently, such justifications tend to be disingenuous and neither especially coherent nor logically consistent. Indeed, if gene pool contamination is the issue, then why target homosexuals if, as presumed, they leave no offspring? More generally, it is hard to sustain an argument that links what is deemed unnatural with that which is ethically wrong. Indeed, counterexamples are easy to find, as much of what we value and consider ethically correct is profoundly—if not deliberately—anti-natural. The point of medicine is indeed to be about as unnatural as it gets: it is our best attempt to stop nature from taking its course, whether it be by curing kids of cancer, taking antibiotics to fend off bacterial infections, swapping out bad hearts for good ones, or overcoming infertility with in vitro fertilization (IVF).

It seems similarly hard to sustain an argument linking some notion of evolutionary imperfection (which is presumably natural, even if error-ridden) to ethical incorrectness. Naturally, there are cases where what seems evolutionarily adaptive is also virtuous—looking after your children would be a case in point. But there are plenty of obverse cases, where things that seem evolutionarily odd are virtuous and things that make great evolutionary sense are anything but virtuous. Perhaps this is best illustrated by the problems of altruism and infanticide. As I mentioned, a core problem for evolutionary biologists is why an organism should be kind to any other at a cost to itself. Indeed, if we all went around giving all our money away, while both evolutionary biologists and economists would be scratching their heads, the ethicists would be applauding, considering this an act of the greatest virtue.

The converse also applies: there is no good reason to suppose that something favored by evolution should be ethically correct. In lions, males often invade a pride, displace the resident males, and kill the cubs (infanticide). This forces the lionesses to reproduce again, to the evolutionary benefit of the invading males. In humans it has also been suggested that infanticide might be more common at the hands of a stepfather, with similar evolutionary rationale given. Even if this is the case, I cannot see how, just because there is an adaptive evolutionary rationale, this in any way can defend the murder of children.

In short, we have cases that are ethically virtuous but evolutionarily problematic, and others that are ethically wrong but make solid evolutionary sense. Arguing for discrimination for or against people based on traits that may or may not have an evolutionary rationale is a nonstarter. Our systems of ethics and laws are there to encourage us, as social organisms, to be civil to each other and virtuous in our actions, and to discourage us from doing our worst. Whether our actions would otherwise have been "natural" or "unnatural," evolutionarily understandable or peculiar, is beside the point. Please, then, don't confuse statements about evolutionary imperfections with statements about the lesser worth of fellow humans. It is no more sensible than to suppose that white moths are ethically of lesser value just because the trees happen to be black. The leap from the assertion of evolutionary imperfection (presumed or otherwise) to stigmatization tells me only about the prejudice of the person making the argument.

Before we delve further into the new problems of our genetics, we need to start by understanding what DNA is and what it does. We can then look at classical explanations for imperfections and see that, indeed, these don't obviously explain the odd nature of our genetics. This is the subject of the next chapter.

2

The Problem of DNA

If there is one molecule that has defined the modern age, it is probably DNA. DNA is the molecule of inheritance of all life. Identical twins are so similar largely because they have the same DNA. The ways in which you resemble your parents are also in no small part usually owing to you having some of their DNA.

DNA was first discovered by the Swiss chemist Friedrich Miescher in the late 1860s. His contemporaries Charles Darwin and Gregor Mendel, the father of genetics, had no inkling of its importance. The chromosome was understood as carrying the heritable material in the 1910s, long before the chemical nature of that material came into focus. In 1944 researchers determined that DNA is that molecule of inheritance, and in 1953 Watson and Crick resolved its structure. This opened the box that led to a thousand discoveries later.

Core to these subsequent discoveries is a set of technological advances that enable us to study DNA. In 1977, Fred Sanger devised a way to determine what makes you different from me at the DNA level, so-called Sanger sequencing. More recently, sequencing technologies have moved on at great pace. The largest sequencing center in the world, BGI in China, currently pushes out 7,000 terabytes of DNA sequence per day. To put that into context, that is 7,000,000,000,000,000 base pairs of DNA every day, while the complete works of Shakespeare contain about 3,500,000 letters.

We understand now how DNA works and how, in turn, genetics works. We use it to understand how cells work, how organisms develop from fertilized egg to adult, and, in turn, to better understand how genetic diseases come about and which diseases we might have. From improved diagnostics we also have improved therapies. Outside of medicine, the understanding of how DNA works has enabled us to determine ancestry on short and long scales: who our parents are (as in paternity testing), where our ancestors come from, and, in longer span, the relationships between species. We use our knowledge of DNA to improve our food and to solve crimes.

But what then is DNA, and in what ways is ours, as a species, odd or apparently imperfect? Can classical explanations for imperfection explain these features? Let me start our adventure by pointing out something that tends to get lost by the wayside: DNA is unusually well suited to large-scale investigation because of its unusual digital biology. The nature of DNA is strange—but very convenient.

The Strange Digital Biology of DNA

While it is good to know that DNA is the molecule of inheritance, this knowledge would be relatively worthless if DNA were hard to study. As suggested above, we are now undergoing an industrial revolution in our ability to study DNA. This convergence of a key molecule and the tools to study it makes for an exceptionally potent branch of science, *genomics*: the study of the full DNA sequence of a species. Genomics is transforming many fields of research, evolutionary biology being no exception. Welcome to evolutionary genomics.

A major reason for the expansion of genomics is that DNA, while a seriously big molecule, is predisposed to relatively simple analysis. To see this, let's look at what DNA is. DNA stands for **d**eoxyribo**n**ucleic **a**cid, but the name doesn't greatly help. Perhaps the best-known feature of DNA is that it is a double helix. Imagine this: take a molecule of DNA by its two ends and untwist it so it is no longer a helix. What have you got? What you will see is a ladder-like structure. The ladder has two vertical rods and multiple rungs. The vertical rods, the phosphate backbone, are the same in all DNA molecules, so we can just omit the analysis of this bit. What is interesting—the variable bits—are the horizontal rungs.

These rungs, however, aren't full-length rungs. Rather, they each are a pair of half rungs that meet in the middle. The half rungs come in four flavors. Let me represent them as four different shapes: a circle, a crescent, a diamond, and a Pac-Man. For those unacquainted with the computer game, Pac-Man was a character with a >-shaped notch for a mouth.

You might see something immediately here, namely that the four shapes I have chosen come in two complementary pairs: a diamond fits into the mouth of the Pac-Man, and a crescent snugly envelops half of the circle. This complementary pairing is crucial to the function of the rungs and of DNA. Imagine, for example, that I take one of the halves of the ladder (one of the two strands of DNA). I could then read off the order of the four types on this one strand from bottom to top of my DNA. Let's suppose that on this one strand these read such that rung 1 is Pac-Man; rung 2, the next one up, is a crescent; rung 3, another crescent; rung 4 a circle; rung 5 a diamond. What is now important, both for how cells work and for how we read DNA, is that, if I know this, I can now also tell you everything about the opposite strand. This must run, from bottom to top: diamond (complement to Pac-Man), circle (complement to crescent), another circle, a crescent (complement to circle), and finally, at position 5, a Pac-Man, complement to the diamond. To analyze DNA, I can ignore both the phosphate backbone and half of the molecule. If I know one half of the ladder, I know the other half.

The same complementarity is key to how DNA gets copied and how two cells, the product of division of one cell, are genetically identical. If you unzip the ladder, exposing the two strands/sides, you could remake full double strands by adding the "partner" of each exposed strand. If on one strand there is a crescent, I complement that with a circle; if a Pac-Man, I patch in the diamond, etc. With the complementarity rule then I will end up with two identical molecules of DNA.

In each of our cells there are 46 chromosomes, two sets of 23: 23 you received from mum, 23 from dad. Each of the 46 is one long molecule of DNA. Our chromosomes have millions of rungs each (very tall ladders!). However, whether the number is 5 million or 50 million, the principle is the same. When a cell divides, each of the 46 DNA molecules is unzipped and then complementarily patched in, giving two copies of each original chromosome—thus 2×46 chromosomes, in effect. The cell then prepares to divide, with each daughter cell receiving exactly one copy of each of the 46 chromosomes. The copying of a DNA molecule to make two identical copies, is, rather logically, called replication.

The four types are in fact complementary shapes, but we don't refer to Pac-Man, crescents, circles, and diamonds. Rather, the four are chemical molecules all of which are chemically *basic* in nature. "Basic" doesn't mean "simple" but indicates an ability to react with acids. The rungs are thus known as "base pairs." These bases combine with a sugar (deoxyribose) and a phosphate to make a *nucleotide*, the full rung including the backbone bit (hence deoxyribo[se]nucleic acid = DNA). The rungs are thus also called "nucleotides," so we also refer to "nucleotide base pairs." The four bases have formal chemical names, which are **a**denine, **c**ytosine, **g**uanine, and **t**hymine. Again, to condense the information in DNA we refer to A, C, G, and T.

The pairing rule is that G and C pair up and A and T pair up. A consequence of base pairing is that all molecules of DNA have as much G as C and as much A as T, since for every G on one strand there is a C on the other and vice versa. Likewise, A and T. This is also why we often characterize DNA in terms of its GC content, meaning the proportion of nucleotide pairs that are GC pairings. While if you know the proportion that are GC pairs you then also know what proportion must be AT pairs (the two proportions sum to 100%), there is no reason to suppose that DNA should have as much A and T on the one hand as it has G and C. Some organisms' DNA is very GC rich (hence AT poor), some very AT rich (hence GC poor). Our DNA is about 41% G or C.

You can then represent a very long, complex molecule of DNA as a string of As, Cs, Ts, and Gs, this reflecting just what is on one of the two strands. It cannot be emphasized how important is this ability to simply—but accurately—represent DNA. It means that when I am sequencing DNA—the term for reading off these bases in order (literally what sequence they are in—I do not have to write out the full chemical structure of DNA. That would be a nightmare! Instead, I can just write out in a very simple text document a series of these letters.

The fact that we can write down a DNA sequence as a sequence of letters captures a further important property of DNA, namely that it is *digital*. Most things we measure in science are not digital, but analog. Temperature, the loudness of sound, and the wavelengths of radio waves, for example, modulate over continually varying scales. Not so DNA—on any strand at one position there is only one of four things you can be. If I asked you to stop somewhere on a staircase it would make no sense to say that you stopped on step 3.678. You can stop on step 3 or 4 but not in between. Similarly, a site in DNA can't be a bit adenineish—it either is or is not adenine. This means that both copying and recording DNA is, at least conceptually, simple. It also means that when you get it wrong you unambiguously get it wrong. This process of getting the copying wrong, or more generally changing one base for another in DNA, is mutation. Because the encoding is digital, once you get it wrong it tends to stay wrong.

This understanding of DNA as a series of rungs on a ladder also gives us a useful measure of the size of DNA molecules. Rather than using the physical length of the molecule if you stretched it out, the size of a DNA molecule is usually measured as the number of base pairs (bp). As DNA molecules are often rather long, to see the numbers better we often measure in thousands of base pairs (kilobases—kb) or millions of base pairs (megabases—Mb). Our 23 different chromosomes we number in approximate order of size, 1 to 22, with the 23rd being the so-called sex chromosomes, X and Y. We thus have two copies of the largest one, chromosome 1, one from mum, one from dad, each of which is about 249,000,000 bp, i.e., 249 Mb. The smallest is chromosome 21 at about 48 Mb. Overall, we have about 3.2 billion base pairs in any one set of the 23 chromosomes, i.e., 3,200 Mb. Incidentally, I suspect I know what you are thinking: Why isn't 22 the smallest if they are ordered in size? When first numbered this was done visually—under a microscope—looking at the chromosomes in a compact form. Chromosome 22 was thought to be smaller, but on sequencing it turned out to be ever so slightly larger than 21 (22 is about 49 Mb).

When first sequencing a species in which all the individuals have two copies of each chromosome (like us), we also often just sequence one of the two sets, as the other set (and other versions of the same chromosome in other individuals in the same species) are usually very similar. If we want to know about DNA variation within a species, we would, however, sequence many different individuals.

Welcome to the Age of Big Data

The sequence of DNA can then be captured in a simple text document, a bit like a very long digitally stored book. Anyone can download one of these for free from global repositories of sequences such as at the National Center for Biotechnology Information (NCBI) in the US. This simple condensing of information has revolutionized the way we can study DNA. Currently we are witnessing an explosion of such information, with the amount of sequence recorded doubling every eight months.

Key to this revolution have been two parallel sorts of advances. The first is clever chemistry to do the sequencing. This is possibly where the biggest changes are happening. It is hard to overstate this transformation. The first sequencing of the 3.2 billion bp of a human genome cost $2.7 billion (about $1 per bp) and more than a decade's work involving laboratories all over the world. This sequence was declared "complete" in 2003. Incidentally, don't take a "complete" genome to mean complete—there are very often troublesome sections of the DNA that are hard to sequence. Our genome was again declared complete in 2022 when these troublesome sections could be patched in. Owing to the advances in sequencing technology, we can now routinely sequence a human genome for between $100 and $1,000 in real time.

This means that patients can, in principle at least, report ill, have all their genome sequenced, diagnosis made, and cure proposed while they wait. Rady Children's Hospital in San Diego currently holds a sort of speed record in this context. They had a five-week-old baby boy present with encephalopathy (a swelling on the brain). With a similar prior history in the family, the doctors suspected a genetic condition, so sequenced his genome. This took a little over 11 hours (not a decade). From this they deduced a particular metabolic problem less than

17 hours after the initial blood sample and 13 hours after they began the sequencing. Successful treatment was started 37.5 hours after admission.

The ability to cheaply, accurately, and rapidly sequence DNA is set to transform many areas of science. The Earth BioGenome Project, which describes itself as a "moon shot for biology," is aiming to sequence all complex life (not bacteria and viruses) within a decade. We also, however, often need information on variation within a species. Rapid sequencing has let us follow in real time mutations in the virus that causes COVID-19 (SARS-CoV-2), which is how we know about new variants very shortly after they first appear. Genetically SARS-CoV-2 is possibly the best-described organism—if we can call it that—there ever has been. The other species we want to understand is us. The US National Institutes of Health (NIH), for example, has a research program, *All of Us*, that seeks to cross-reference over a million people's DNA with their medical records.

The second great technology leap concerns our ability to handle and make sense of this data. With the amount of DNA data doubling every eight months, for this we need large computers and computer programming abilities. We also need appropriately skilled people. A word of career advice: if you like computers, you stand a great chance of getting a well-paid job these days being a so-called bioinformatician, one of the folks who computationally analyze all this data.

Extracting Meaning from Data

The big data problem is not just about storing all those strings of nucleotides. Just reading into a computer a string of As, Cs, Ts, and Gs is not itself all that helpful. We need to extract meaning from the sequences. Key to that is understanding what DNA does.

A core function of DNA is to encode the sequence of amino acids that make up proteins. Proteins are the molecules that do much of the heavy lifting in cells, be it enzymes to digest food to make energy, structural molecules to allow our muscles to contract, or the proteins to enable replication. Proteins are involved in every cellular process. They can be many thousands of amino acids long, but on average, they tend to be about 500–1,000 amino acids long. To understand DNA, we need thus to understand which proteins it codes for.

Just as DNA is a linear sequence of nucleotide base pairs, proteins are a linear sequence of amino acids. The linear string of amino acids will fold up into all sorts of shapes, but they are all fundamentally linear strings. Indeed, you can think of proteins as being a bit like balls of string, except some are long and thin, some more ball-like. The fact that both DNA and proteins are linear series of units, rather than, for example, branching chemical structures like some storage sugars such as glycogen, provides a clue as to how the sequence of nucleotides specifies proteins. Step one, called *transcription*, is a bit like replication. A bit of DNA is unzipped, and the sequence of nucleotides on one strand is used to make a single-stranded complementary molecule. In this case, this new complementary molecule is then released from the unzipped DNA and the DNA zips back up again. Information in the DNA—the sequence of bases—has thus physically been transferred to this new, single-stranded nucleic acid, called RNA (ribonucleic acid).

The odd part of the process is what then happens to this RNA in step two. It goes to a special small machine in the cell, the *ribosome*, and there the information copied from the As, Cs, Ts, and Gs of DNA is translated into a different string, this time of amino acids. For this reason, this process, from one coding system to another, is called *translation*. As it is just an information intermediary between the DNA and the resulting protein, this RNA is thus called *messenger RNA*, or mRNA for short. The rules of translation are now well established. The message in the RNA is read in non-overlapping blocks of three (a block of three is called a *codon*), each block specifying a particular amino acid or a signal to stop translation. The length of the protein in terms of numbers of amino acids is thus one-third of the length of the coding sequence in the mRNA, as three base pairs in the RNA give one amino acid. The rules for reading the blocks of three comprise the so-called *genetic code*. Much like Morse code, where . . . means S, - - - means O, and so on, the codon ATG, for example, is translated as the amino acid methionine, and GGG specifies the amino acid glycine.

Physically what happens is that there is an adapter molecule that both recognizes the specific codon in the mRNA and has attached to it the corresponding amino acid. The adapter is *transfer RNA*, or tRNA for short. As each codon is read by the ribosome, the amino acid specified (attached to the tRNA) is chemically glued to the one before it in the string. Thus, a gene that starts ATG GGG would result in the first two in a string of amino acids being methionine (ATG) and then glycine (GGG). A linear string of nucleotides thus specifies a linear string of amino acids.

There are in total 20 different amino acids that we use that collectively give proteins their different properties. With 4 different bases, and 3 bases in a codon, you might then have noticed that this gives 64 different possible codons ($4 \times 4 \times 4$). An mRNA terminates at one of three so-called *stop codons* (TGA, TAA, or TAG), leaving 61 codons that code for amino acids. With just 20 amino acids, this means that several codons can all specify the same amino acid. GGC, GGT, and GGA, for example, also specify glycine. As it happens, methionine is coded by ATG alone.

Whether there is any rhyme and reason to rules relating codons to amino acids is one thing I used to study (it isn't as random as it first seems), but that is a story for another day. What is relevant here is that, if we know the genetic code of a species, i.e., the set of rules linking codons to amino acids (most organisms use the same "universal" code), and we know the DNA of this species, and with similar technology can work out the mRNAs, then we can also, with a computer, determine where the protein-coding genes are in the DNA and what proteins they code for. The key step is that we can map the sequence in the mRNA back to the DNA because the two are complementary.

To a text document with the series of DNA nucleotides, we can thus add an "annotation" of the genes of the genome. That sort of information was vital to working out the metabolic disorder of the baby whose DNA was sequenced. The team could identify a section of DNA in the gene coding a particular protein that was altered from how most of us have it, and that was very likely to have caused a badly functioning protein of that gene and hence the disease. If you are interested, the gene goes by the name *SLC19A3*, and the protein sits in the membrane that surrounds our cells and allows thiamine (vitamin B1) to go from outside of cells to the inside.

While I am talking about the genetic code, let me briefly get something off my chest. You will often see in the press something like "Scientists have discovered the genetic code of such and such a species." I get a bit hot under the collar when I see this. You can nearly always be sure that this they haven't done. The "genetic code"—the set of rules—is nearly always the same for all species. What is usually being reported is a *whole-genome DNA sequence* of the new species. You have been warned.

New Data, New Problems

It is perhaps inevitable that, with all this new data, new intellectual problems would also arise. One such concerns that most basic of properties: How many genes do we have? A stated goal of the Human Genome Project was to determine the location of the estimated 100,000 human genes. This sounds pretty definitive: we have about 100,000 genes. In 1990 this was estimated down a bit to the 50,000–100,000 range. Shortly before the genome was completed, an "estimate the number" competition was launched: the GeneSweep contest. The first bets were taken over drinks during an annual genetics meeting, all hoping to win the $3,000 prize. Of the more than a thousand entries, estimates ranged from about 26,000 to more than 312,000. The average was near 40,000.

We still are not sure of the actual answer. Given that there are also mutations of a form that either add or remove DNA, so we can all be born with more or less DNA than our parents, there may well also be no single correct answer. The two first "complete" genomes suggested about 31,000 and 26,588 protein-coding genes. This was subsequently revised down to 22,287. Current best estimates are lower still, ranging from 19,901 to 21,306. The very best current estimate, then, is that we have about 20,000 protein-coding genes. Not only is that lower than anyone guessed, it is about half the average estimate. However, we should probably not exaggerate this issue too much, for two reasons. First, many years prior to the GeneSweep contest there were several estimates in the 20,000–30,000 range. It is a bit misleading to suppose (as many do) that the number was completely unexpected (but this field tends to have a rather short memory). It does seem fair to say that the number is at the very low end of expectations. Second, we are not very good at deciding whether a short gene is real or not. There could be lots of small proteins that we have yet to count.

Leaving this issue to one side, to put that 20,000 into context, single-celled yeast that we use in brewing and bread making has about 6,000 genes, the laboratory fly *Drosophila* has about 14,000, the laboratory worm *Caenorhabditis elegans*—a millimeter-long thing with very little in the way of complexity—has about 18,000, while our best-studied plant, the tiny weed *Arabidopsis*, has about 25,000; microscopic water fleas have about 31,000, rice plants probably have around 51,000, and a tiny single-celled organism that causes a common infection in industrialized countries, *Trichomonas vaginalis*, has an estimated 98,000. Why then do we have so few protein-coding genes, fewer than we expected and many fewer than organisms that by any standard are much less complex than us?

This low number brings with it a further surprise. While the classical view of DNA is that its function is to specify mRNA to specify proteins, only about 1.2% of our DNA specifies the proteins we need (fig. 2.1). By contrast, we now know a lot about the DNA of very many bacteria, and they could hardly be more different. Their DNA is almost all (typically about 85%–90%) coding for protein. And if it isn't coding for protein, it is most commonly the bit of DNA (that sits just in front of genes) that acts as part of the on/off switch controlling when the gene will be transcribed. Why do we have these two extremes?

Most of our DNA, then—about 98.8%—does not specify strings of amino acids. We can roughly divide this other DNA into two sorts— DNA between genes, and non-coding DNA within genes. There is a bit of wobble here, as we now know we have about the same number of "non-coding" genes as protein-coding genes. These non-coding genes are sections of DNA that are transcribed, making RNA, but not then translated. It remains an unresolved problem how many of these are doing something important or are just so much transcriptional noise.

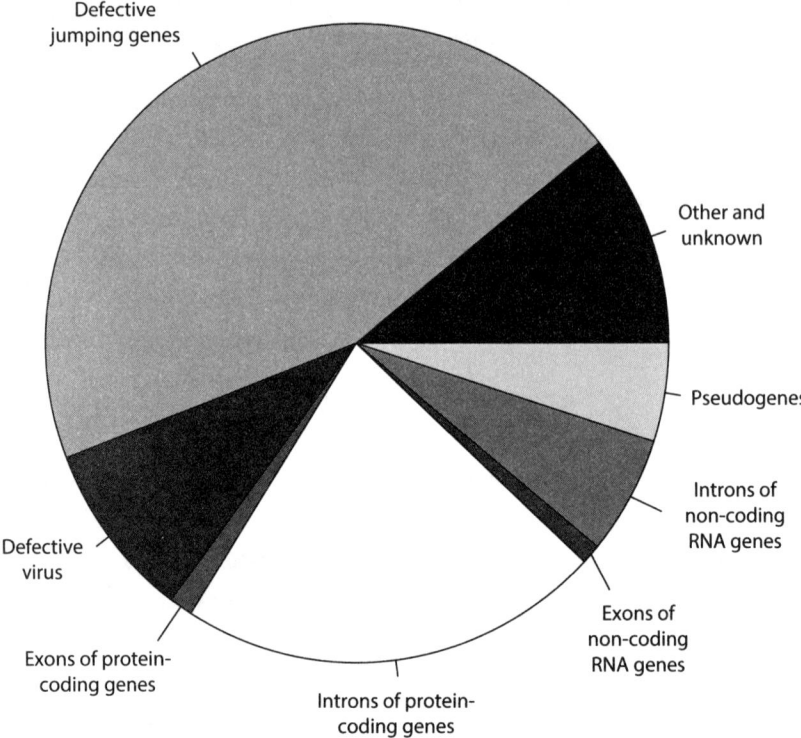

FIGURE 2.1. **The composition of the human genome.** Very little of our DNA codes for protein. Most is old defective jumping genes (transposable elements) or remnants of viruses that inserted into our DNA. Included in "other" are about 0.1% live jumping genes, 0.1% live viruses, regulatory sequences concerned with turning genes on and off (about 0.2%), as well as specialized sequences at the ends of chromosomes (telomeres, also 0.1%) or at chromosome pinch-points necessary for chromosome segregation during cell division (centromeres: 1%). Numbers from https://sandwalk.blogspot.com/2018/03/whats-in-your -genome-pie-chart.html.

We also have some genes sitting within the non-coding bits of other genes, like Russian dolls inside Russian dolls.

But what is this non-coding DNA within our genes? To understand this, we need to return to the processes of transcription and translation, as I skipped over an important part of how our mRNAs (but not those of bacteria) are usually made. Remember that from DNA we make a single-strand copy, the RNA. You can think of RNA as a bit like a large reel of the sort of tape first used to record films. For most of our genes, this "immature" RNA is then cut up into chunks, some bits removed, and the rest spliced back together in sequential order (in the technical jargon we use this movie editing term, "splicing"). The bits excluded—the *introns*—are usually sent to a cell recycling bin. The others, the *exons*, are spliced back together to make the mRNA. It is these that tell the cell what protein to make. I must at this point apologize for our language here. You might have thought that the bits of RNA that we *ex*clude are called exons and the bits we *in*clude in the mature mRNA the introns. Sorry, no, it is the other way around. Go figure.

Leaving the confusing etymology aside, the very fact that we have introns is odd. While bacteria have no introns like these, we have high numbers, and they are very big: over 90% of a typical human gene is non-coding intron that must be cut out. Consequently, over a quarter of our DNA is protein-coding genes; it just so happens that most of the RNA transcribed from each gene is removed before making its protein. Why does so much of our protein-coding genes not code for protein?

Looking across species, as we now can do, the mystery deepens. For any genome, we can ask how much DNA is between genes, what the average size of introns is, and how many introns there are. These give three independent measures of what we could call genomic compaction or, conversely, bloating. A compact genome would have few introns within its genes, the introns it did have would on average be small, and little DNA would sit between genes. A bloated genome, conversely, would have lots of introns that on average would be large, as would the length of DNA between genes.

Given that these three measures are independent, there is no logical reason why you should not have, for example, a large gap between genes but small introns within genes. Looking from single-celled organisms to plants, animals, and the like, we find, however, that these three measures are correlated: species that have lots of introns tend also to have bigger introns, and the intergenic DNA tends also to be longer (fig. 2.2). The number of introns in intron-containing genes is the weakest predictor, but the relationship between intergene spacer size and average intron size is striking.

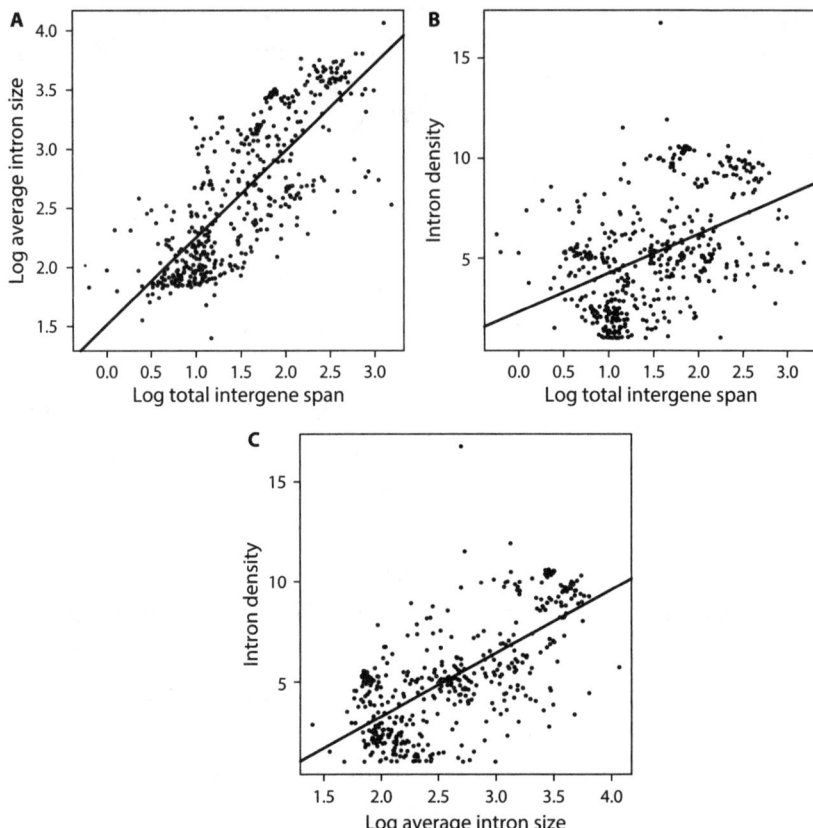

FIGURE 2.2. **The relationships between three independent genomic "size" variables—log of the average intron size, intron density, and log of the total intergene span—for over 400 eukaryotic species.** Average intron size is the mean size of introns; intergene span is the total span of DNA between coding genes; intron density is the mean number of introns in genes that have introns. Data from: Lozada-Chavez, I., Stadler, P. F., and Prohaska, S. J. 2018. "Genome-wide features of introns are evolutionary decoupled among themselves and from genome size throughout Eukarya." https://www.biorxiv.org/content/10.1101/283549v1.article-info.

Compare us to baker's yeast, for example (one of our very best-understood genomes). Few of their genes (about 5%) have introns, whereas only about 5% of our genes *don't* have at least one intron. On average, we have about 10 introns per gene. If one of their genes has introns, it usually only has one. The average size of our introns, at about

5,500 bp, is much longer than in yeast, where introns are about 200 bp. Our longest intron, in a gene called *ROBO2*, comes in at an enormous 1,160,411 bp. The average size of each of yeast's 16 chromosomes, each with about 300–400 genes, is less than that, at about 750,000 bp. Not unsurprisingly, our between-gene DNA is much longer too. For us, there is probably about 10 kb on average between genes; in yeast it's more like 1 kb. Why then do we have lots of mysterious DNA, and why do all three independent measures of genome bloating (amount of between-gene DNA, number of introns per gene, and average size of introns) seem to evolve in the same way?

There are many other oddities of our genome that make us stop and ponder. For example, for most of our genes we cut the "reel of tape," the immature RNA, in many different ways—many different splice forms. Many, if not most, of these different spliced forms are then sent to a cell's recycling center before doing anything. They often appear to be faulty, with a stop codon in the wrong place (and we have a system to detect mRNA with stop codons in the wrong place). So why make them in the first place? We also see that our cells make all sorts of what look like errors in making the reel of tape and in processing it after it is spliced. It all looks rather poorly made. Why?

At a more focused level we can also analyze stop codon usage. I will explain. Each protein-coding gene needs, at the end of the mRNA, one of three "stop" codons that instruct the machinery to stop making protein. These are, as I mentioned, TGA, TAA, or TAG. We know these aren't equally good. TAA is relatively effective, meaning that the machinery almost always stops at the stop sign. TGA, however, is not so good—it is like a rather faint red stop sign that can easily be missed, leading to cellular traffic accidents. The best estimate is that it gets it wrong about every 1 in 500 times, while TAA gets it wrong about once every 2,000 times. Not surprisingly, then, natural selection favors usage of TAA in most organisms. Mammals, however, are odd—we overuse and conserve the least good one (TGA). Why?

Perhaps the strangest of all parts of our biology is how bad we are at making babies. Surely after millions of years we would at least have gotten this right! Our ability to study DNA has allowed us to understand what the problem is: in most cases, it is because of mistakes in handling DNA. Look at just-conceived human embryos (what are referred to as "early" embryos) and we see that many have the wrong number of chromosomes (the rate of these errors is lower in young mothers, higher in older ones). We canonically have 46 chromosomes—23 from mum, 23 from dad—but very commonly these early embryos have 47 or 45. One IVF study from 2014 estimated that of 284 embryos, 151 (53%) had abnormalities in the number of chromosomes. This estimate may be a bit on the high side, as other estimates suggest slightly lower numbers for non-IVF embryos. Nonetheless, these mistakes are still remarkably common, affecting between 40% and 60% of all embryos. One study on human eggs before fertilization reports that 70% have the wrong number of chromosomes.

Nearly all these chromosomally abnormal embryos die very early, before the mother knows she is pregnant. There are thus many human conceptions that the mother knows nothing about. But we also have nonsensically high rates of miscarriage, in which the mother knows she is pregnant and loses the baby. A study of over 120,000 pregnancies between 1978 and 1992 in Denmark, for example, estimates that 13.5% of known pregnancies result in known miscarriage. Other estimates suggest 10%–20%. If we merge the IVF study with this Danish study, then, through their lives, women have 1.7 live births on average and 2.1 pregnancies that didn't survive. Similarly, for Mormon women living in nineteenth-century America the numbers are more like 8 live births to 17 that didn't live. For humans, pregnancy failure, commonly owing to the wrong number of chromosomes, is the norm, with 60% or so of all conceptions failing.

And it gets stranger. From the sequence of the DNA, we can work out which parent the extra chromosome came from. You might imagine that half the time it comes from mum, half the time from dad. But no, it is nearly always from mum.

It doesn't have to be like this. If we look in lowly yeast, which like us can reproduce sexually, we see the same sort of problem, except there the rate of error is massively lower: about 1 in 100,000 fertilizations are faulty. Perhaps the high rate of chromosomal errors in early embryos is not a feature of simpler organisms but is a feature of all complex verte-brates? The data suggest not. In our vertebrate relatives—fish, frogs, birds—rates are also extremely low. Their fertilized eggs also have extremely low initial mortality rates. In fish in the lab, for example, there is almost no embryonic death (by any cause) prior to the depletion of the yolk, while humans have very high death rates immediately post-fertilization. In other mammals, rates of having the wrong number of chromosomes are also high, as in humans. In flies we can also estimate the number of eggs that die or lead to developmental disorders. In one study, from over 5,000 eggs, 10% or so died before becoming larvae, we think a few percent owing to the wrong number of chromosomes issue, and none of those that survived had a developmental anomaly.

Why, then, do we make so many embryos with the wrong number of chromosomes?

How to Explain Imperfection?

The more we study DNA, the more oddities we see—things that look enormously like they might be imperfections. How might we explain these? The problem of imperfection is not new, and the field of evolu-tionary biology has provided a diverse series of explanations. Many are interrelated, but let's try to break them down. These we can broadly characterize as: It takes time for evolution to catch up with a changing environment; we aren't in the environment we evolved in; evolution can only tinker with what is there; it depends where you start from; and the world of possibilities is limited. Let me try to explain each.

It Takes Time

The first problem gets to the heart of how evolution works. Evolution-ary change involves one variant replacing another variant over many generations. Sometimes this happens in real time, for example when the Omicron variant of SARS-CoV-2 displaced the Delta variant (which displaced the Alpha variant, etc.). Nonetheless, it isn't instantaneous. In this case it was fast because Omicron has a very short generation time and is effective at spreading in populations (it is very infectious). You could imagine an alternative way of changing: for example, Apple makes a new update to their iPhone operating system and everyone with an iPhone updates as soon as this is available (this is more or less what happens). In principle, the operating system of all iPhones could be changed immediately and in parallel. Mechanistically, this is not how evolution works. Organisms don't decide that there is a new and better version of their species' DNA and all suck it up, overwriting what was already there. It is much more similar to competition between operating systems, with one slowly outcompeting others in the market: currently this is mostly Android versus iOS, with older mobile phone systems (e.g., from Microsoft and Nokia) having been driven to extinction.

As generation times tend to be rather long (about twenty years for us, compared to minutes or hours for bacteria and viruses), we usually think of evolution as a slow and gradual change (this being implicit in the notion of evolution not revolution). Being slow and gradual, however, has consequences. Imagine you went back to the early nineteenth century and discovered that some moths were black, but most were white, while the tree bark was all covered in industrial soot. The moth seems far from perfect—why aren't all members of the species camouflaged? The answer is simple: it takes time. All members of all species don't change immediately. There are some amazing species, such as chameleons, that can instantaneously change their color to match their background, but most camouflage isn't like that—it is genetically hardwired. Each moth can be black or white, but, unlike chameleons, not both. Consequently, for these hardwired species, we need the slow replacement of one type by the other.

Many suggest that this sort of process explains much about us. For example, we haven't evolved to stand up all that long ago, and so for some of us our knees and backs are not as evolutionarily brilliant as they might be. Humans indeed more commonly suffer from bad backs than do our primate relatives. The stress on the back from walking upright is thought to partially explain this. However, we are not all equally prone to bad backs. For example, look at the bone region in our back between the spinal bone at the bottom of the chest area (the last thoracic vertebra) and the top one of the lower back area (the first lumbar vertebra) and you will see that we aren't all quite the same. Some of us have a thing called Schmorl's nodes, some don't. These are depressions found on the upper and lower surfaces of the bones. For those of us lucky enough not to get a "slipped disk" (alias intervertebral disc herniation), this region is clearly different from that of chimps. For those unlucky enough to get slipped disks (those with Schmorl's nodes), there is no big difference between their structure and that of chimps. For evolution to change a species from having one state to all having a better state takes time. And boy, do some of us know it.

Walking upright is a peculiarly human thing, but evolutionary lag is expected to affect all species when they are adapting to parasites (any other organism that can infect us and be harmful, be they viral or bacterial, intestinal worm or fungal foe). This is because parasites have short generation times, so spread fast and evolve fast. For example, some people are resistant to HIV via mutations in a gene, *CCR5*. In places on the planet where HIV is common, these mutations are increasing in frequency in the local population, much as black moths increase in frequency over time as white moths tend to die early. But it takes time for all individuals in a population to have such resistance, just as it took many generations for a population of white moths to evolve into a predominantly black population.

Right Body, Wrong Environment

A closely related sort of explanation is that we aren't adapted to the current world. If you take a population of black moths, well adapted to sooty tree bark, and suddenly move them to a new, cleaner environment, they will look out of place, literally. Environmental mismatch can make for new problems and new imperfections as well. The human lineage diverged from the common ancestor with chimps about 6 million years ago, and nearly all that time our ancestors lived in Africa. Only very recently (in the last 80,000 years or so) did the ancestors of modern humans migrate out of Africa and colonize the planet. As a species, we have spent very little time not in Africa.

Despite this, we see some evidence that some populations have adapted to their new local environments. For example, Tibetans, Ethiopians, and natives of the high Andes have all evolved to be able to live in the thin air at high altitude. This is not a simple physiological response that comes about if any individual were to live at high altitude for a long time (the aim of high-altitude training in athletes). Rather, this is owing to the spread in these populations of mutations in genes that enable their blood to grab what little oxygen is available. In two of the populations the same gene (*EGLN1*) has independently been under selection to enable this. Because of this evolution, Tibetan babies have, on average, higher birth weights, higher oxygen saturation, and are much likelier to survive than other babies, such as those of the recently established Chinese population, born in this environment. These Chinese are just not adapted to their new environment, as they haven't lived there long enough to genetically adapt to it. The Tibetans hold a speed record for the time to adapt—we think it may have taken about 3,000 years. However, this figure is debated, as it doesn't make much sense of the archaeology, and 5,000–6,000 years may be more likely. Either way, for human populations to adapt to their environment takes time. Recent movement (in the last few thousand years) to a new environment will mean we are out of whack, genetically speaking.

A bigger problem may then be that we aren't moving to a new environment, but that our environment is changing instead. I'm not thinking about climate change here. Rather, in the very recent past our exposure to microbes and the food we eat has changed.

Only very recently have we had the option of a sugar-rich Coke and fat-loaded Big Mac as a meal. Some argue that because in our evolutionary past energy-rich sugar and fat were both nutritionally important and rare, we evolved a craving for both. This possibly explains why our junk food is Big Mac and fries, not salad and nuts. In turn, the argument goes, we can now be victims of preferences that served us well in our old environment but are harmful in the modern world. It is hard to be sure whether this model is fully correct. We can have a good idea of what people ate in the past, and a high-fat, high-sugar, high-salt diet would have been unusual. It is, however, hard to know about selection on dietary preferences in the past. It makes sense that we might have cravings for things we need but that are hard to find. But making sense and being true are not the same thing. What we can be sure of is that a high-sugar, high-salt, high-fat diet isn't doing us any good. It leads to obesity, heart disease, cancer, and type 2 diabetes.

Perhaps one of the best case histories has been described in Micronesia, a province of the Pacific Islands. Shortly after World War II the US military surveyed the population and found no obesity, high blood pressure, or diabetes. In the 1960s and '70s, the islands received food aid from the US, and the diet transitioned to rice and imported foods. By the 1980s there was an increase in sugar and sweet foods. More recently the diet has moved to rice, wheat, sugar, refined foods, and fatty meats. Obesity rates have gone up: three-quarters of adults are now overweight or obese, and a quarter of the adults have type 2 diabetes. If we ask why we are imperfect in getting heart attacks, diabetes, and cancer, then here is part of your answer. We just are not adapted to current food availability.

A more subtle effect may come about because of our recent ability to keep our world spotless and to avoid getting dirty. With a decrease in infectious diseases has come an increase in allergies and so-called autoimmune disorders (conditions in which our body's immune system attacks itself, type 1 diabetes being an example). The "hygiene hypothesis" holds that this is a consequence of living in a sanitized world that we never evolved in. Specifically, it is suggested that some infectious agents (parasitic worms, bacteria, etc.) that co-evolved with us might be able to protect against a diverse collection of immune-related disorders. Take them away and the immune system misfires, overreacting to a trigger, be that peanuts or pollen, that should be treated as immunological friend, not foe.

The first evidence for this came from David Strachan of the London School of Hygiene and Tropical Medicine in 1989. He was interested in hay fever, whose prevalence was enigmatically increasing through much of the twentieth century in the UK. At the same time, asthma and childhood eczema were also becoming more common. He found that hay fever was least common in children with lots of older brothers and sisters. Why might that be? Although he wasn't sure, he suggested that unhygienic contact with your (unclean) older siblings could explain this. The trend for smaller families and greater hygiene, he speculated, might thus explain why these diseases are becoming more prevalent.

Since the turn of the twenty-first century, it is estimated that about 1 in every 5 children in industrialized countries suffers from an allergic condition of some flavor, be it asthma, hay fever, or skin rashes, all of which increased in prevalence over time. Since the 1950s rates have approximately tripled. That the same increase is seen for type 1 diabetes (which is also hitting children younger) is an important part of the evidence. Type 1 diabetes is like the allergic conditions in that it is a case of a misfiring immune system. In the case of allergies, we misfire against external agents (like pollen), whereas in type 1 diabetes our immune system misfires against our own proteins (so-called autoimmunity). Whenever we see increases in this disease or that disease, we always first suspect better diagnostics. However, as type 1 diabetes is straightforward to diagnose, the increase and earlier age of first onset are unlikely to be simply explained by such a bias.

The hygiene hypothesis (in some form) has much going for it. Globally, areas with higher levels of allergic and autoimmune disease are those with low levels of infectious diseases such as gastrointestinal infections. We can be reasonably confident that this doesn't have all that much to do with genetics, as type 1 diabetes is sixfold higher in Finland than in neighboring Karelian Russia, whose people come from the same genetic background.

Perhaps the best evidence that the higher levels of allergies and autoimmune disorders in the West have something to do with the environment is that offspring of immigrants coming from a country with a low incidence of type 1 diabetes have as high a rate as their new country within the first generation. This has been shown, for example for people who moved from Pakistan to the UK. Perhaps this has nothing to do with increased hygiene and everything to do with higher air pollution, or less sunlight and associated lower vitamin D levels? Air pollution almost certainly affects some conditions. Indeed, asthma levels were lower in the less polluted West Germany than in the old East before the Berlin

Wall came down. That we see an East–West tendency across Europe, with lower rates of type 1 diabetes in Bulgaria and Romania than in Western Europe, argues against a sunlight-mediated effect. The rates in Bulgaria and Romania are going up rapidly since the fall of the Wall.

We also see some evidence that infection is protective. Owning a pet and exposure to farms—especially cowsheds—seems to protect against eczema, especially if the mother visits when she is pregnant. Giving people infections such as an intestinal helminth worm (*Trichuris suis*) has been found to decrease symptoms in patients with Crohn's disease (a gut problem associated with immune system involvement). Long-term gut worm infections also cause a rise in an immune defense molecule, interleukin-10, low levels of which are associated with asthma. Conversely, treatment to get rid of gut parasites like helminth worms can result in increased eczema-like conditions.

By no means is all evidence supportive. For example, deworming treatment sometimes results in reduced asthma; using worms to treat inflammatory gut diseases doesn't always work; measles seems to be associated with an increased risk of allergic disease; and COVID is associated with increased risk of childhood type 1 diabetes. The hygiene hypothesis has thus morphed into a new idea: the "old friends" hypothesis. This model suggests that it is the picking up of harmless microorganisms, the "old friends" that have been present throughout human evolution, that is key. These somehow train the immune system to react appropriately to threats. For our immune system it isn't just knowing what to attack that is important; it also needs to learn what to tolerate.

There certainly are bacteria that we need, for example in our guts, where they help us digest food. It would be a mistake to mount an immune response to rid us of our wanted friends. There exists a possible mechanism underpinning the old friends hypothesis, namely that recently discovered regulatory T cells in our immune system are there to damp down immune responses and that these cells are stimulated by our old friends. Fewer old friends then lead to an over-reactive immune system that doesn't know that pollen and peanuts are harmless.

The implications of the old friends hypothesis are different from the hygiene hypothesis. Both agree that exposure to a diversity of microbes is a good idea. However, the old friends hypothesis would not recommend getting a nasty disease, like measles or COVID, as the cure. Rather, it is particular non-disease microbes that we need (so don't stop washing your hands).

An area of particular interest is whether Cesarean sections might indirectly cause allergies and immune disorders. During a C-section the baby is removed by surgery from the womb and isn't exposed to all the microbes in the vagina. The suggestion then is that the increased rates of asthma, allergies, and immune deficiencies in people delivered by C-section (this seems to be the case) might be owing to lack of exposure to the friendly vaginal microbes. Tests are being done to see if seeding babies with vaginal juices might mimic vaginal delivery and to see whether C-section babies and vaginally delivered babies have different microbes. Given the complexity of the assembly of microbes in all of us, it is perhaps not surprising that the answers to the second question are mixed. It may be that C-section is important not because of maternal vaginal microbes but because of gut ones instead. The utility of vaginal seeding is unresolved, it may be harmful, and it is currently not recommended.

Evolution the Tinkerer

No matter whether it is the hygiene hypothesis or the old friends hypothesis, there seems to be a good body of evidence that the modern Western world has reduced the diversity of microbes that we have exposure to, and that this goes some way to explain increased rates of immune overreacting conditions. As such, these imperfections are likely to be a consequence of us very recently inhabiting an environment we didn't evolve in.

We also did not evolve to live in warm, well-insulated houses with running water and electrical appliances. Not so long ago I lived in a lovely early-eighteenth-century house. While there are many good things to be said about "period charm" (as the estate agents/realtors would put it), the electrics and the plumbing are not one of them. There were pipes and cables in all sorts of odd places, running down walls, across the ceiling. There were many pipes that came down through the floor above that, to this day, I have no idea about. If you ask why it was like this (as many old European houses are), the answer is pretty easy to see: these houses weren't built when there was plumbing or electrical wiring, so the house was never designed with them in mind. The wiring and plumbing were put in long after. Usually this meant that the pipes and cables are not where they would be, were you to design the house and build it today. I presently live in a modern house and never see pipes or wiring anywhere.

The evolutionary process is a bit like making changes to my old house. Changes add to what is there, they don't start from the ground up and redesign everything. Evolution is a tinkerer. This can indeed explain many of the oddities of our plumbing and wiring, too.

We have "wiring" in the form of nerves that, like copper electrical wires, conduct electricity—the parallels are quite striking. Consider then the strangeness of the left recurrent laryngeal nerve. The vagus nerve starts in the head (near the bottom of your brain), then runs down your neck. A branch of this, the left recurrent nerve, then runs under the artery that takes blood away from the heart (the aorta) and back up the neck again. It thus runs parallel, and in the opposite direction, to the parental vagus nerve—this is how it got its name "recurrent" (in anatomy-speak this means moving in the opposite direction). It eventually ends up close to its point of origin, supplying electrical impulses to our voice box (larynx). This structure describes a very long U shape, whereas it could have been only a few inches long at most had it taken the direct route from a to b. All mammals have this same quirk. In giraffes this makes for a detour of almost five meters.

The reason for this is that species are not all new houses built to a perfect plan. Evolution just plays with what is available. The nerve is also seen in fish (and other vertebrates). In fish it goes from the brain, past the heart, off to the gills. In fish, there is no detour. More or less, the nerve could run over the aorta or under it and the distance would be about the same. It happened to evolve to go under. You could think of development of the nerve as being given directions to put into the GPS, but rather than from a to b, the instructions are to go from a to b to c: brain to heart to gills. In us, as early developing embryos, when these structures are being developed (in the womb), we use the same rules as the fish do. Our developing nerve has the old instructions to go south under the aorta and then turn north. Over evolutionary time, however, the neck has extended, and the heart moved lower into the chest. Over evolutionary time the distance from *a* to *b* and *b* to *c* has become vastly greater than the distance from *a* to *c*. For this reason, it just seems plain silly. Especially if you are a giraffe.

Just as necks elongated through evolution, so too testes went from being internal to being external. This has left a similarly odd bit of construction, this time in our plumbing. The tube (vas deferens) connecting the testes to the prostate gland (just under the bladder) runs over the tube (ureter) connecting the kidney to the bladder. The vas deferens is, consequently, longer than it needs to be. It could have gone from testes to prostate and be done with the loop around the ureter. However, in organisms in which testes are internal (e.g., fish, reptiles, birds, etc.), going under or over the ureter makes little to no difference. In our evolutionary history, with the instructions built in, the routing instructions stayed the same, and another strange detour is the odd result. Incidentally, the dropping of the testes over evolutionary time explains another of our imperfections: our tendency to get hernias. These happen when parts of the gut extend through the muscle lining of the gut cavity. The muscle lining needs to have the ability to open to let the testes through, and it is this resulting weakness that causes hernias—the gut pops out through the same openings through which, in males, nerves run.

If You Want to Get There, Don't Start from Here

Evolution is a tinkerer in two senses. Not only can it only start with what is available, it also tends to make small, incremental changes. To see why this might lead to apparent imperfections, let's play a game. I want you to go mountain walking and to get as high as is possible. Perfection, in my game, will be standing on the top of Mount Everest, as this is the world's highest mountain. But I'm going to make life a bit harder. I am going to blindfold you (so you cannot decide which is the highest peak you can see), insist that you start wherever you currently find yourself (no jetting off to Nepal), and say that, while you are free to pick any direction to walk in, you can never go downhill. These restrictions match the restrictions of evolution by natural selection. The restriction to only go uphill is like having the better replace the less fit. You can't decide where to go because evolution doesn't have a plan. You also need to start where you are because evolution only plays with what is currently available. Mutation is random, so will go in any direction.

Will you ever get to the top of Everest? Almost certainly not. Depending on where you start, you may get no farther than a tiny little hillock. Evolution by natural selection is rather similar: it just enables species to climb the fitness peak that they happen to be at the foot of. In the technical language, we even talk about "hilly" fitness landscapes and the "peak-jumping" problem. Species not only evolve from other species, but they almost always only explore, by mutation, possibilities really similar to the current state. Humans don't suddenly give birth to cats.

We see plenty of evidence for this sort of problem in evolution. There are, for example, over thirty occasions when eyes of one form or another have evolved independently. But they seem to have climbed mountains of different heights. If we look at our eye, for example, we see that the cells that sense light at the back of the eye are sort of back to front.

You can think of an eye as being a bit like a digital camera in which light comes in through a small hole at the front, is focused using a lens, and is projected onto the back of the camera/eye. Individual receptors then detect whether they have been hit by light and send this information, via electrical wires, to a computer that processes this information. The wires are neurons that conduct electricity. The computer for us is the brain. As it happens, rather curiously, the part of the brain that interprets the signals from the eye, the visual cortex, is about as far from the eye as possible at the back of the brain. The signal from the left eye also goes to the right side of the brain and vice versa. All very weird, but I digress.

The oddity within the eye itself is the anatomical arrangement of the electrical wires from the receptors to the processing unit. The sensible thing to do would be to run these wires out behind the receptors (as you see in digital cameras). But we don't. Our receptors are back to front, and our cabling runs out from the receptors toward the front of the eye, before all the wires are bundled together. The bundle then runs through the back of the eye off to the brain. This gives us a blind spot in our eye (the fovea) where the bundle passes through the back of the eye.

Importantly for the present question in hand, not all eyes do this. The eye of the octopus is effectively the same as ours (one lens, receptors at the back of the eye, a bundle of electrical cables taking the output of the receptors to a processor). In this case however, the wiring runs out behind the receptors so there is no blind spot. It looks like when octopuses evolved eyes, they climbed a higher mountain than we did.

The World of Possibilities Is Limited

One solution to the peak-jumping problem (how to get from a good solution to a better solution without going downhill) is that occasionally evolution may not be a tinkerer. Perhaps sometimes mutations make really big changes, big jumps from mountain to mountain. In flies, because of the way their development is programmed, you get seriously weird mutants that have legs where there should be antennae (the gene controlling this got the name "Antennapedia," *pedia* meaning legs or feet). Thus, indeed, big mutations can sometimes be seen, especially in development. In the 1930s there was speculation that evolution may happen because of such "hopeful monsters." Big-effect mutations certainly happen, but whether they ever underpin evolutionary change—as opposed to making weird flies in the lab—isn't so clear. What is clearer is that the rules of development in any species can influence the possible forms that can be produced. This can better explain what we don't see than what we do see.

In mammals, for example, we have never seen in the wild a baby born that didn't have a father. We see fatherless babies in birds, fish, amphibians, etc., just not in mammals. They aren't common in birds, but in some fish and reptile species, such asexual reproduction is the normal model of reproduction.

We think we know why mammals are odd. We have a strange way we transcribe about 100–200 genes in early embryos called *genomic imprinting*, not seen in these other species. While we have two copies of all our genes on the 22 non-sex chromosomes, for some we use only the one we inherited from our mother or only the one we inherited from our father. When you were an embryo in your mother, the gene for insulin-like growth factor 2, for example, was transcribed exclusively off the DNA you inherited from your dad. If any of the genes transcribed just from paternal DNA code for vital proteins, then embryos without father's copies will be unable to develop. Hence, we never see babies without fathers. This was, indeed, the big problem in trying to clone mammals, Dolly the sheep being the first success. Cloning toads had been worked out much earlier and is, by contrast to cloning mammals, pretty straightforward.

The idea that the way development works prevents certain states from being realized is sometimes called *developmental constraint*. Note that it is different from an explanation that, for example, asexuality is possible but selection favors sexuality. That would be just normal natural selection.

There is a similar developmental rule thought to explain another weirdness: that different species of centipedes all have an odd number of trunk segments. The number of segments is variable between species, from 15 to 191. But the number is always odd. This could be because selection strongly prefers odd numbers. Or it could be that because of the way development is wired, mutation is likely only to ever create progeny with odd numbers. It looks like the latter is true. In centipedes' development, segments come in blocks of two. Mutations that change the number of segments do so by changing the number of these blocks and so increase the number of segments by even numbers. One head block is, however, an exception—it has one unit. Thus, centipedes start with an odd number of segments. As they can only go up or down in twos, they all have odd numbers of blocks. Even if there were some advantages to having an even number of segments, selection cannot favor a trait if mutation cannot generate it.

In other cases, the world of possibilities isn't limited by what can be created, but by what can be created and survive. This may explain—somehow—a strange pattern. Have you ever looked at the skeleton of one of those long-necked dinosaurs? Compare that to the neck of a giraffe and you will see something odd. Mammals, from tiny shrews to long-necked giraffes, nearly all have the same number of neck bones—seven (there are a few odd exceptions such as sloths and manatees). To make a long neck, mammals increase the size of their neck bones. Dinosaurs (along with birds and other reptiles), by contrast, increase the number of neck bones.

Why do mammals not usually alter neck size by altering the number of vertebrae? The best answer is that, unlike even-numbered segments in centipedes or asexuality in mammals, it is possible, but babies with extra vertebrae die owing to immediate knock-on effects. Specifically, there is an increased risk of cancer in the babies, stillbirths, and neural problems. Exactly why is not fully clear, but the same genes that control the development of the vertebral column also contribute to cancer. The suggestion, then, is that if a mutation enables an extra neck bone, that same mutation also causes cancer. Why reptiles and birds aren't affected is also not quite so clear; however, they do seem less prone to cancer.

This would be one example of a more general idea that in biological systems it can be hard to change one thing without changing another at the same time. This can be because genes and their proteins have many different jobs, or the same job but in many tissues. For example, in humans there is a gene that makes a protein that converts one amino acid (phenylalanine) into another (tyrosine). Mutations in this gene prevent this from happening and cause the disease phenylketonuria. As the conversion is important in many tissues, the result is intellectual disability, pigment defects, and eczema. Similarly, we see in chickens a mutation that makes their feathers curly (the gene is called *Frizzled*) that also affects body temperature, digestive capacity, and egg-laying ability. In humans and cats, we see mutations that affect both hearing and pigmentation.

This idea—that if you change one thing another part of the system also changes—has a fancy name: *pleiotropy*. It is also a rather slippery term. In medicine, for example, it refers to the idea that a mutation can result in different diseases in different people. Evolutionary biologists tend instead to think that it would mean that it causes more than one effect in the same individual.

Pleiotropy might put the brakes on evolution in more-complex species. The more complex the species—the more different parts, you might say—the greater the chance that a mutation could benefit the organism by affecting one of the parts but also be harmful because it affects others. This is sometimes known as the cost of complexity. However, it remains unclear how important this is. Jianzhi Zhang and colleagues at the University of Michigan compiled data on many mutations in yeast, worms, and mice and, for the most part found that most mutations affect just one trait.

The evolutionary importance of pleiotropy rather depends on why one mutation affects many traits. For example, sometimes pleiotropy can be a consequence of nothing more than the laws of physics, rather than anything special about the way genes work. With limited resources to make offspring, you can have many small kids or a few large kids, but you cannot have many large ones. There is nothing particularly biological about this. If I gave you a small amount of money to buy shares on the stock market, you could buy a few expensive shares or many cheap ones, but not many expensive ones. These are so-called trade-offs.

You could say that not having large numbers of large babies is an imperfection. If it is, it isn't one evolution can do anything about. When, however, the pleiotropy is owing to contingencies of the way the system happens to be wired, then there is at least the possibility that the system could be rewired.

Perhaps one of the most interesting areas of imperfection where these sorts of effects play out is aging—or, more accurately, *senescence*, the decline in performance with age. A leading explanation for this is that you cannot be both brilliant when young and brilliant when old, just as you can't have many big babies. Indeed, in fruit flies those with high fecundity early on also live a shorter time.

Beyond such pleiotropy, important to the evolution of senescence is the notion that selection cannot act to favor mutations associated with increased performance when we are old. This is because, owing to the causes of death beyond our control—so-called extrinsic mortality—few people live to be old. A mutation, for example, that enables bearers to have 20:20 vision at age 200 would not be favored over one that enables 20:20 vision at age 150, as no one lives to, or past, 150. More generally, mutations that only affect—for good or bad—the old (post-reproductive) are invisible to selection. Mutations that benefit the young but harm the old, however, can be favored.

How then would selection shape the rate at which species age? For a species like us, as we get older, and by chance will die for some external reason, we are best off living fairly fast and dying fairly young. The calculation is different in different species. There is an Australian marsupial mouse-like critter, *Antechinus*, that, probably owing to scarcity of food during the dry season, has no prospect of surviving to reproduce the next year after having reproduced this year. Individuals of this species go through an aging process very much like ours, but, unlike ours, theirs is in real time only hours or days after they are done reproducing. In males this means after copulation, for females it means after weaning. They live fast and die really fast.

Conversely, there is a debate as to whether organisms like large coni-fer trees age in this sense at all. Naturally they get older—each year they add a tree ring. However, when they survive to reproduce the next year, they can have even more offspring (via seeds) than they did in the year before just because of the way they grow (in a tree-like manner, you might say). The same equation favors them to keep on investing in stay-ing alive, and so they don't obviously decay with age—or they do so much more slowly than we do. Look at a big pine tree when next you see one. It is remarkable how the old ones often look amazingly healthy.

No matter whether we talk of developmental constraints, pleiotropy, or trade-offs, the core idea is the same: there are some states that might be perfect that you cannot achieve because the change just isn't possi-ble, or isn't possible without other consequences. All the trauma we endure in getting old may well be the best selection can do, given that we can't have our cake and eat it too.

So Why Is Our DNA So Odd?

We have then a rich suite of explanations for the evolution of imper-fection. Indeed, examined in the round, we as a species aren't doing so badly—many of our apparent imperfections are better seen as components of alternative ways of organizing a life course, the different solutions found by each species being the best each species can be (different "life history strategies," in the official jargon). Compared to *Antechinus* we have a long lifespan, and our three score years and ten are associated with gradual decline (senescence), not the real-time decay of this micro-marsupial. They never have the luxury of time to moan about bad backs and dodgy knees. Conversely, compared to giant redwoods we seem rather less good by such measures.

By some metrics, humans, and mammals generally, are in fact some of the most efficient reproducers. As populations are on the average neither expanding nor contracting in numbers, each mating on average produces two progeny that survive to reproduce. A salmon making 5,000 eggs (about their average) must then see 4,998 of these never make it to be parents themselves. Imagine what would happen if for every salmon in every generation, even half the fertilized eggs survived—the planet would be drowning in salmon in no time! Our comparable rate of mortality prior to mating (i.e., the proportion not surviving to reproduce) has to be much lower, not least because over her lifetime a human female probably produces no more than about 500 eggs. In terms of mortality rates *per unit time*, species like fruit flies must have a high rate, as they don't live long. In any given fly generation, all fruit flies will be dead long before the time it took you or me to finish weaning. If measured in these terms, we again look like a pretty good species. Fast-living species with a vast excess of offspring must then have much higher mortality rates than us, by all measures.

My aim is not, however, to produce a beauty pageant of organisms and to measure each against some benchmark of perfection. This would overlook the truism that each species is great given its circumstances (extrinsic mortality rate, bodily anatomy, etc.). Indeed, if we use the number of offspring surviving to reproduce as a metric of evolutionary brilliance, then to a good first approximation all species are the same. More than two survivors and the species concerned would be forever increasing in numbers—which cannot happen; less than that, and it would be heading to extinction—in which case it most probably isn't a living species. It is a question of different strokes for different folks, you might say. *Antechinus* does just fine in its ecology, giant redwoods in theirs. It is interesting to understand their biology and how different life histories evolve, but in the end, it is a bit pointless to say that one species is any better than another.

Are humans then really so imperfect? Whether one wants to call our peculiarities "imperfections" is rather immaterial. What is more important is that there are features that, even in the above context, still demand an answer and suggest that our understanding of the evolutionary process isn't complete.

When we look at the evolution of humans, and mammals more generally, there are aspects of our biology—most especially our genetics—that suggest that evolution has gone backward. The March of Progress has turned around.

Most other species overuse the best stop codon, while we overuse the worst. Most of our DNA isn't protein-coding, while most of that of our ancestors (far enough back in time), by contrast, is obviously functional. We make many embryos with the wrong number of chromosomes, but our close(ish) relatives, fish, do not. As we shall see, we also have one of the highest rates of mutation, and with that, a high rate of genetic diseases.

The problem of our inability to properly make fertilized embryos is all the more baffling when compared with what is seen elsewhere. Indeed, would it not make more sense if salmon, with 4,998 of 5,000 eggs dying, had very high rates at which they made eggs with genetic defects, as most will die anyway? But they don't have high rates of eggs with the wrong number of chromosomes, and we do. Our rate of harmful mutations also seems to be higher than theirs.

Naturally, this leaves the problem of what does prevent the 4,998 salmon eggs from getting to mate. Given the lower rates of mutation and chromosomal handling errors in fish, whatever the cause, it isn't obviously genetic. In zebrafish held in the lab, mortality peaks at the point when they need to become independent feeders: if they don't get a meal they die. The embryos are otherwise fine.

These funny features of our genetics can't then be easily dismissed as imperfections in name only, just an alternative strategy optimal for our ecology. The better states are attainable—just not by us.

They also aren't obviously explained by the sorts of explanation we saw above. They can't be chalked up to being a consequence of developmental constraints, since the mechanisms by which we make sex cells, process genes, copy our DNA, etc., are not obviously different from those in other species. They don't obviously have anything to do with being in the wrong environment, not least because the same problems beset other mammals. The problems associated with peak-jumping don't seem to apply, not least because, in terms of this model, we seem to be walking *down*hill. Far from being the pinnacle of the evolutionary tree, we seem to be somewhat retrograde. While species that produce huge numbers of progeny of which only a few survive persist *because* of their life history strategy, we survive *despite* these features.

We seem therefore to be missing some aspect of the evolutionary process. What follows is my attempt to synthesize why we—and mammals more generally—are like this. It largely boils down to two things: being physically large and having a womb. Big-bodied species have small populations, and in small populations the perfecting force that is natural selection is severely blunted. Having a womb, and constantly feeding our young in the womb and after, opens the way for all sorts of problems, including—counterintuitively—making it beneficial for some genes to mess up how we make sex cells. Seen from this vantage point, it is remarkable that any of us are alive.

Before we go there, we need to look again at the problem of how natural selection works. When we understand that, we can in turn understand both why it is not the only important evolutionary process and why it sometimes doesn't lead to organisms being better than their ancestors. That is the subject of my next chapter.

3

To Begin at
the Beginning . . .
How Adaptation Works

In some of my favorite days away from the desk, I go into my local schools to explain the joy of understanding evolution. Inspired by the peppered moth, we sometimes set the younger age groups a task: with tweezers in hand, they pretend to be birds pecking at moths. The moths in this case are paper cut-outs, some black, some white, put onto black paper. The students count the moths before they peck away, and after. They discover that after their predation the relative frequency of black moths has gone up.

Having explained that in the nineteenth century, industrial pollution turned the trees black, I ask: How was it, did they think, that the moths also turned black over time? A not uncommon response is that, just as the trees became covered in soot, so too the moths. Their answer is, in this instance, wrong, but logically brilliant. Indeed, if trees go black because of soot, why would moths not do the same?

On further probing, we discover that the students understand perfectly well that the black moths are less likely to be eaten (they understand selection as a filter), but they don't necessarily mechanistically couple this with the increase in frequency of the black type, relative to the white morph, over the longer term.

What is missing from their understanding is *inheritance*. Indeed, the task we set them—while well-intentioned—may reinforce their logic: How could paper cut-outs reproduce? They are just bits of paper! To understand the process of natural selection, we need to better understand a sometimes-hidden component of the story: the genetics.

So much do we emphasize the logic of selection that we often forget to mention that selection alone is not sufficient for the process of evolution by natural selection. Indeed, if moth color were not heritable, natural selection could not be a possible explanation for changes in the relative frequency of the black and white types. If every time moths mated there was no difference in the proportion of black and white offspring, then no matter what selection did in one generation (perhaps only the black types survived), on mating the proportions would be reset to the same start each new generation. A lack of heritability would be evidenced by white mated with white, black with black, and black with white all giving the same proportion of black and white. If this was so, there could then be no increase in relative frequency over the longer time span owing to natural selection. We would have to think about moths getting covered in soot as the best explanation.

In many regards, the principles at stake are not especially biological. If, like me, you follow your favorite sports team (I have the misfortune of being a lifelong Leeds United supporter), you may have noticed that the position of teams in a league table after one season is correlated with their positions in the prior season. In English football (soccer in the US), over the last few seasons the top of the Premier League is some combination of the same teams. Currently these are Manchester City, Manchester United, Liverpool, and Chelsea. Arsenal and Spurs are usually up there too (Leeds are not). The reason these teams repeatedly do well is not simply owing to their wealth as such, but because quality begets quality. Part of this is because the teams and management are largely unchanged between any two seasons. The wealthiest teams hold on to the best players. More generally, in having greater wealth they can both hold on to and also recruit the best players—this allows season-to-season inheritance of quality, even with some degree of player movement.

Imagine what would happen if, instead, each season all players and coaches were randomly allocated to teams. The club's wealth would be largely irrelevant, as the richest have no means to hold on to or attract the best players. There would then be no reason that last season's best football club would be this season's best. There would be no season-by-season inheritance of player quality. Consequently, the team that did really well this season (the equivalent of black moths on a black background) would have progeny (next season's team) that had no greater resemblance to the winning combination than any other team. Competition—in both biological and sporting terms—will not lead to longer-term dominance without some sort of carryover of quality. In biology this carryover is mediated by DNA.

While in outline season-to-season player carryover and generation-to-generation DNA inheritance have comparable properties, there are also some peculiarities that follow from the strangeness of DNA. Built into the genetics are limitations on what selection can, and cannot, achieve. To see this, let's start by thinking about the gradual process of step-by-step adaptation.

Step, by Step, by Step

Understanding the gradual evolution of camouflage is as good a way of understanding numerous aspects of the odd process of evolution as any. Let's think about a species like the peppered moth. Moths are odd. They fly at night and rest during the day. But resting during the day is a dangerous thing to do, as it is during the day that your predators can see you. Unless you can hide away, looking like the background you rest on seems like a pretty good idea. Camouflage is thus a good solution to the moth's predicament, but let's rewind the clock to a point where the background and the moth were not concordant. I know: let's suppose the tree trunks where you reside are mottled white and you are black (hmm . . . wonder where I got that idea from).

In the first instance I'll assume all moths are genetically identical for their color traits and are all a gray-black. We start by rolling the mutational dice. Mutations hit DNA all the time at a low rate. To be a heritable mutation the mutational process needs to happen in cells leading to sperm or eggs in the prior generation—in the *germline*. If this new mutation is transmitted to the son or daughter, it will be found in every cell of these offspring. As individual parent moths each transmit one of their two copies of their genome to each of their progeny, we will say that each individual has two copies of any given gene. Using this metric, in the first generation there would have been one copy of the mutant gene in the population: just one individual will have the mutant gene, and only one of its two versions of the gene will be the mutant.

Let's imagine that random mutation generates a variant that is even more black, and that, because it is a germline mutation, this could be inherited (the offspring are also even more black). If moths with this version stand out more on the white mottled background, we expect this mutation to be eliminated from the population, as moths with the new mutation are more likely to be eaten. This one mutant version of the color gene appears by mutation, but leaves the population because of selection. This is known as "purifying" or negative selection. The population reverts back to everyone being the same gray-black. Death before mating because of a mutation that induces a genetic disease is the same process.

Now imagine a different germline mutation. This again is some change to the DNA that, we shall suppose, causes a little white mottling. A moth with this mutation doesn't greatly resemble the background, but resembles it better than everyone else. What will happen to this? If this increase in mottling means that the moth with this new mutation is less likely to be eaten, then, on average, we expect to see it at a slightly higher frequency in the next generation. The mottled moth we assume has two copies of all genes (one from mum, one from dad); it stands a slightly higher chance of surviving to reproduce, and when the moth reproduces, the new mutation has a 50:50 chance of going to any given offspring, owing to the way organisms transmit chromosomes. Imagine that all mating pairs have the same number of offspring. Overall, then, our mutation goes up slightly in frequency in the first generation because it stood a higher chance of being in a body that survives to reproduce, not because it is inherited by more offspring per reproduction.

At this point it is worth clarifying what we mean by "frequency," as it can be confusing. Frequency here is defined not in terms of the absolute numbers, but in terms of proportions: it isn't the total number of mottling-causing genes in the population—this is the *absolute* frequency—but the *proportion* of all the alternative versions of that color gene (black vs. mottled) that are mottled. If there are N individuals in the population, each has two copies of the black/mottling gene— one from mum, one from dad. There are thus 2N copies of the gene. The frequency of the mottling mutation is just the proportion of these 2N that have the mutation—the *relative* frequency. When we talk of the "frequency" of a mutation in the population we nearly always mean this relative frequency, not the absolute frequency.

There is a reason for this. When a mutation first appears in the population by the process of mutation, there will be one bit of DNA with that mutation, so its relative frequency at the start is 1/2N; its absolute frequency is 1. In part, the reason we focus on relative frequency is that we are interested in evolution defined as a mutation going from rare to being the only version of that section of DNA in a population— frequency 2N/2N =1. We aren't so much interested in the number of individuals in a population. Changes in the number of individuals isn't evolution. Put differently, if there are 10 copies of a mutant gene in a population of 100 individuals, this population grows to 1,000, and the number of mutant copies is now 100, there has been no evolution in our eyes: the relative frequency is the same. The other reason we focus on relative frequencies is that it also makes the math much easier.

Back to the plot. What happens in the second generation? Perhaps in the second generation there are now two individuals with the new mutation, not the original version. The total number of copies of this gene (in mutant or original form) is the number of individuals in the moth population times two, because each moth has two copies of each gene. If there were a million moths of this species, the new mutation would have been at a (relative) frequency of 1 in 2 million, but then, because its bearer stands a better chance of being one of the moths that doesn't get eaten, it is now a bit more common, 2 in 2 million. This contrasts with purifying selection, where the mutation also is initially rare in the population but becomes even more uncommon, indeed absent.

Fast forward and the beneficial mutation will go up and up in frequency simply because bodies it is in are less likely to be eaten: from 2 in 2 million to perhaps 4 in 2 million, and so on. Give it long enough (many generations) and all the population will have it: 2 million out of 2 million copies of this gene are now what was once the new, rare version. In technical language we say that the mutation is now at *fixation*, or that it has "fixed." Each time a mutation goes from rare to common like this we talk of a *selective sweep*. A selective sweep takes time—many, many generations.

And so now we have a population in which all individuals are a little mottled. That was sweep number one. The moths in this population still don't look much like the background, but they look more like the background than the original moth. This captures an important principle about natural selection: what matters is how competitive you are relative to others, not in absolute terms. You might alternatively think of this as "in the land of the blind, the one-eyed man is king." You only need to outperform the competition—you don't have to be perfect from the get-go. The same is true of sports teams: you don't need to be perfect, you only need to be better than the opposition. There is a similar joke to the effect that when being chased by a lion you don't have to be able to run faster than the lion, you just must be able to run faster than others being chased by the lion. For this reason, evolutionary biologists compute in terms of "relative fitness" (how good am I compared with my competition), not "absolute fitness" (how good am I compared to some perfect state) when trying to determine the fate of mutations.

This same first bout of positive selection (also sometimes called Darwinian selection) holds another truth. In this genetic process the transformation of the population is not sudden. It isn't that all individuals somehow figured out what the better DNA looked like and either magically changed their DNA to look like the better version, or somehow sucked it up. Nor should the fact that the process improved the population's survival (by reducing the chances of being eaten) be taken to mean that mutation is in any sense directed. There are plenty of mutations that reduce the chances of survival (negative or purifying selection), and some may well have no effect. The selection that results in a change to the population is due to a filtering process that acts on new mutations.

Incidentally, you might sometimes see claims that evolution can't lead to improvement because it is a random process. This is just a strange misunderstanding: mutation is random, selection is a directional filter acting on that variation. It is no odder that a population changes over time to better fit with its environment than that a colander lets the water through (bad mutations being eliminated) and keeps the pasta in (good mutations being captured).

If that was round one, then what happens next? Individuals in a slightly mottled population will keep randomly mutating. Mutation isn't triggered by the first mutation getting to fixation. Mutation happens all the time, just at a very low rate. Again, some mutations will be harmful to the bearer (going back to more black, less well camouflaged), some of no consequence, and some will provide a slightly better mottling with a better match to the background (at least in the eyes of the predator species). When a mutation of the first type appears, we expect it to be eliminated from the population (purifying selection), and the population will revert to everyone being a little mottled. When the third type appears, it may well again go from 1 in 2 million to 2 in 2 million, etc., not because of repeated mutation, but because bearers of the mutation are again a little better off than the rest of the population. A second "selective sweep" sees this new version now go to fixation after many generations.

This process can keep repeating. Each bout would be natural selection; the full set of repeated bouts is the process of *adaptation*. The term "adaptation," by the way, should come with a warning. Sometimes you see people refer to each bout of natural selection as "adaptation" (even the professionals do this). Sometimes the much longer-term series of selective sweep after selective sweep is the process of adaptation—as I defined it above. And sometimes the word is used to denote the end product, rather than the process. The well-camouflaged mottling of the moth is "an adaptation" and the process by which it came about is "the process of adaptation." Ugh. I do wish our language was better. (Incidentally, there is no such word as "adaption." It is "adaptation.")

The Great Evolutionary Dartboard

You might reasonably ask: Will the process ever stop? In principle, if you ever got to perfection, it should stop, as at this point no new mutation can improve the fit to the environment. In this context we can think of evolution as being like a dartboard, with each throw (new mutation) having to be closer to the bull's-eye than the previous one that fixed, for it in turn to become fixed. As, over time, after multiple sweeps, the population gets closer to the center, eventually a mutation that is any better may be impossible, in which case you have reached the limit of perfection.

The dartboard model has some interesting properties. First, it tells us why most mutations are bad for you—and cause genetic diseases, for example. At the limit, as we get closer to the bull's-eye, the successful mutations will be ever finer modifications, and there is more room for mutations to be harmful—hitting anywhere further away from the bull's-eye. When organisms are close to perfection, random mutations are more likely to take you further away from the best state, especially big-effect ones. There is nothing biological about this: if you randomly fiddle with the mechanism of a beautiful Swiss watch, you are similarly more likely to make the watch worse. Consequently, mutations should mostly make mechanisms worse.

Second, it suggests that mutations that make big changes are unlikely to be important, except possibly right at the start of the process. Imagine the population starts a long way from perfection. We could say that any mutation that so much as hits the dartboard will be better than the current version of the gene. Let's suppose the first bout of the process of adaptation involves such a throw. It hits the dartboard, let's say, in the outer ring somewhere (like our initial weakly mottled moth mutation). This mutation arrives and runs to fixation. We are still far from the bull's-eye. The second dart (second set of mutational events after the fixation of the first beneficial mutation) now has a more limited space to land to be an improvement, as the successful dart needs to be closer to the bull's-eye than the first one. And so on and so forth. The effect size of the successful mutations we thus expect to get smaller and smaller as we get closer and closer to perfection. Sculptors can similarly take away big chunks of marble when starting a sculpture, but make ever finer strokes as the sculpture nears completion.

This suggests that big-effect mutations can only really be expected to be successful when a population is far from perfection. At least for simple organisms, there is some good evidence for this. One of the most important recent efforts to understand how evolution works has been done by Rich Lenski and colleagues at Michigan State University. He initiated the so-called long-term evolution experiment (LTEE). In this he has been tracking genetic changes in twelve initially identical populations of the bacterium *Escherichia coli* (*E. coli* for short). The experiment started in February 1988 and has been going on ever since. It reached 73,000 generations in early 2020 but was frozen (literally) during the COVID pandemic. It started again in September 2020. The starting bacteria were not well adapted to life in the laboratory. One feature Lenski found in all twelve lines was that improvement in fitness, mediated by genetic changes causing faster growth rates and increased cell size, was initially rapid but slowed thereafter, just as the dartboard model predicted.

They also, however, found some utterly unexpected massive changes in just one of the lines. All the bacterial lines were grown in a medium containing normal bacterial food as well as chemicals *E. coli* doesn't eat when oxygen is present, including citrate. Indeed, a defining feature of *E. coli* is that it cannot use citrate in the presence of oxygen. Much to the researchers' surprise, one day they discovered that one culture of bacteria had gone cloudy. This lineage of bacteria had suddenly found mutations that let them eat the abundant citrate. So big-effect mutations do exist and can thrive. This example, however, also adds complexity to the simple story. On checking back by looking at the ancestors of this line (they froze samples all along to be able to do this), they found that the selection on the mutation allowing citrate consumption was much more likely to happen on some genetic backgrounds than others. So the simple story of mutations filtered by selection isn't the complete story. Chance events in the past can also be important.

Whether the dartboard model simply translates in the same way to more-complex organisms isn't so clear. This dartboard model of evolution (very similar to the hill-climbing metaphor we saw before) can be thought of as a simple manifestation of the British evolutionary biologist and statistician R. A. Fisher's so-called geometric view of evolution. Fisher, by the way, was both a genius and a man of what now would be considered dubious opinions. He was an early eugenicist and in favor of voluntary sterilization. He also advocated that smart folks like him should have more kids—and practiced what he preached. He was also deeply opposed to the idea that smoking causes cancer. Even the smartest folks can't be right all the time.

Fisher's model is commonly thought of as what you might call a multidimensional dartboard, each dimension being a different trait. In this more complex version, some mutations, if of big effect for one property, may, especially in a multicellular species, also be bad for some other property (this being the pleiotropy mentioned in chapter 2). This logic underpins the problem of the costs of complexity, the idea being that more-complex species have more interwoven genetics, meaning that any one mutation is likely to affect many different properties. As big-effect mutations are unlikely to be successful (they might move you closer to one bull's-eye but further from another), we might expect evolution to proceed more via small-effect ones. There is an ongoing debate about just how much this affects the process, but it may mean that the first mutations cannot be as big as they might have been in less complicated organisms.

Selection in Action

The logic above seems clear, but can we see this in action? With sudden shifts in conditions—like industrial pollution—we can see beneficial mutations sweeping through populations. The rapid spread of the citrate metabolism in Lenski's bacteria is another example of a sweep. However, perhaps one of the most fully described examples of selection in action, including understanding of what genes are involved and how, is evidenced in a deer mouse (*Peromyscus maniculatus*).

In this species there seems to have been recent evolution of coat color driven by bird predation. Hopi Hoekstra of Harvard University and her colleagues were especially interested in these mice in the Sandhills of Nebraska. These dunes formed 8,000–10,000 years ago and are made of light-colored quartz. The surrounding older lands are dark-colored. Because they are geologically young (yes, 10,000 years is young) and make a distinct ecology, any evolution of the deer mice is expected also to be recent. One such feature of the mice is, you might have guessed, coat color. If you catch mice on these various habitats you find, not surprisingly, that the darker the soil, the darker the coat color. These coat color differences are associated with mutations in one particular gene, *Agouti*. This gene mediates the production of a yellow pigment, pheomelanin.

This all looks like what you would expect given selection. But can we see evidence for selection?

To address this, Hoekstra and her colleagues set up an experiment. Bite marks on plasticine mice had told them that the most common cause of predation was from birds. To see if bird predation could select for or against some coat color, they set out six 50 × 50-meter pens that stopped ground-dwelling animal predation but permitted birds to attack. Three of these were on dark soils and three on light soils. Each pen had no resident mice, but equal numbers of light and dark mice were then introduced. The team then tracked the populations over a few months to see what happened to the relative frequencies of the light and dark types. As you might expect, lighter-colored mice taken from light soils put into light soil pens did better than the dark mice. Likewise, dark mice did better on dark soils. This is selection.

What was going on genetically? In the *Agouti* gene, as they expected, they found mutations that had changed in frequency during the experiments. One was especially interesting. This was a change at one amino acid: some mice had a serine amino acid at one place in the protein and some mice were missing this (mutations don't always swap one nucleotide for another, they sometimes cause loss or gain of DNA as well). This absence or presence was strongly correlated with the light or dark coat color in the original populations living on light or dark soils, and also showed a big difference from the starting frequency in the experiments to the end frequency. This mutation was a strong candidate for at least some of the changes in color seen over time.

What did this presence or absence of a serine do? The team could do some clever genetics: take a different mouse species that naturally lacks the *Agouti* gene and make new mice with that gene engineered in, some with the *Agouti* missing the serine, some with the version with the serine. The ones missing the serine were indeed lighter than the ones with the serine included. This matched what they saw in the selection pen experiments, where the selection for or against light coat color corresponded to increases or decreases, respectively, in the relative frequency of the missing serine.

But what does this missing serine do? Looking at the hairs of the mice missing the serine, they found that they were lighter-colored because they had reduced amounts of the pigment pheomelanin, the one controlled by *Agouti*. They could go further. The missing serine is in a part of the Agouti protein that binds another protein called *attractin*. This binding of the two proteins (Agouti and attractin) is needed for pheomelanin production. They could show that the version with the serine binds more strongly than the version missing the serine. The mutation causing a loss of serine thus causes weaker attractin binding, which causes lower pheomelanin production, which causes the mice to be lighter-colored. In the pens with dark soils, this missing-serine version went down in relative frequency, most probably because birds can see light mice against a dark background more easily than dark mice against a dark background. In two of the three light-soil pens, the version of Agouti with the missing serine went up in relative frequency, but the effect in light-soil was not as dramatic as it was in dark-soil pens.

While this analysis of coat color in deer mice provides evidence for selection and what exactly it acts on, it is like one snapshot of evolution. Can we see experimental evidence for multiple bouts of sweep after sweep? As evolution tends to go faster in species that reproduce faster, the best way to look at this is using bacteria or viruses. We have seen sweep after sweep in SARS-CoV-2, the virus behind COVID, now one of the best-studied case histories of evolution in action we have. The original variant became common, this was replaced by the Alpha variant that was replaced by Delta that was replaced by Omicron (version 1, BA.1), which was replaced by Omicron version 2 (BA.2). We also see sweep after sweep in Lenski's *E. coli* population, with cell size, for example, getting larger in each of the lines in each sweep.

An especially visually striking example of multi-bout evolution has been provided by Roy Kishony of Harvard Medical School and colleagues. Usually when doing work with bacteria in the lab we put bacteria in small round petri dishes. The petri dish has some jelly-like substance added to it for the bacteria to consume. The first antibiotic was indeed discovered by Alexander Fleming using such a setup. He saw that bacterial growth on a petri dish was inhibited in the vicinity of a fungal growth on the same plate. The fungus, he discovered, was releasing penicillin, which was killing the bacteria.

However, we also now know that if one leaves such a setup long enough, bacteria will often evolve resistance to the antibiotic. This is the same process as described above—the bacteria are mutating all of the time, and very occasionally a mutation appears that enables a bacterium to survive better in the presence of the antibiotic. What Kishony and colleagues did that was different was to do the same sort of evolution experiment, but this time in a massive dish, one that was 1.2 meters by 60 centimeters. Rather wittily, they called this a "**m**icrobial **e**volution and **g**rowth **a**rena," or MEGA, plate. The plate had bands running up and down it of different concentrations of an antibiotic, increasing as you move to the middle of the plate. Motile bacteria were then introduced at the ends of the plate, which were free of antibiotic (this way you could run the experiment in parallel—one from each end). This first population depletes the local bacterial food supply, thus selecting for any mutants that can move to the next band with a higher concentration of antibiotic.

What they observed was rather beautiful. At the leading edge of the first band there indeed appears by mutation a bacterium that can resist the low-dose antibiotic, and its progeny spread within this band (you can see them fanning out from one point). They then stop when they get to the higher-dose next band. But here again, one of this group of bacteria finds a genetic solution to this higher dose and now spreads into the new band. And so on and so on, sweep after sweep.

This experimental setup had a further peculiarity. If a bacterium mutates to be resistant to the antibiotic but isn't physically near the leading edge of the band, it doesn't invade the next band—bacteria on the plate don't jump about, they just move locally. Indeed, Kishony and colleagues found mutants that would have been good but were trapped behind the leading edge. As with the deer mice, selection has a spatial aspect in this case, too.

Right Here, Right Now

This simple view of random mutations coupled with a selective sieve has many implications. The first is that evolution is not a process with foresight. We all use foresight: we check the weather forecast before leaving the house and try to guess what clothes to pack for our holidays. Animals have foresight too: squirrels and nuthatches, for example, bury seeds for future use. Trees dropping leaves before winter storms can be thought of as foresight. While natural selection can favor individuals in species to have such foresight, it cannot favor traits that are currently bad for you but will be good in the more distant future.

The advantage or disadvantage that any mutation has is dependent just on the conditions individuals face in their life spans. It doesn't act now on conditions sometime in the distant past or the distant future. There is no moth that evolved to become black because it somehow expected that in nineteenth-century industrial England this would be a good idea. Selection just cannot do that. The mortality happens because of today's birds seeing a moth on the current background.

In retrospect this seems like a bit of an obvious insight. It also explains some apparent imperfections. Consider, for example the need for vitamin C (ascorbic acid) in our diet. For British sailors on long sea voyages, vitamin C deficiency and the resultant scurvy was a major issue, solved by having lemon juice added to the daily ration of watered-down rum. Incidentally, this remedy is sometimes claimed to be the result of one of the first recorded experimental trials. In 1747, James Lind, a ship's surgeon in the British Royal Navy, gave some crew members two oranges and one lemon per day, while the "control" group were given something else (cider, vinegar, sulfuric acid [!], or seawater), in addition to normal rations. The citrus supplement group uniquely avoided scurvy. However, as always, the story is not so "neat": the British Admiralty was slow on the uptake. Their more systematic employment of lemon juice was made almost fifty years later (in 1795), largely independent of any advice from Lind. In no small part this was also Lind's fault. While his small trial found what looks like good evidence for the consumption of oranges and lemons, nonetheless he proceeded to confuse matters by recommending alternatives, such as malt, and thought the main cause of scurvy was not diet but confinement. Such is the rocky road to medical progress. Ascorbic acid also got that name because people were looking for the anti-scurvy factor, or the anti-scorbutic factor, to be more exact (a-scorbutic became ascorbic). Its chemical name is really L-hexuronic acid.

But why can't we synthesize vitamin C for ourselves? Is this not an imperfection? Indeed, in most mammals, vitamin C can be synthesized, so it isn't a dietary necessity. If most other mammals became sailors, they would not need to pack oranges, lemons, or limes (which is why the British are sometimes known as limeys, by the way). You want some fancy chemical names concerning how vitamin C is synthesized? Well, vitamin C comes from L-gulonolactone, which is produced from L-gulonate, which comes from D-glucuronate, which comes from uridine diphosphate glucuronate. The names really don't matter. What matters is that each step is enabled by a protein enzyme encoded by a specific gene. The most interesting of these in the present context is the only one uniquely involved in vitamin C production. It is the one doing the final step, converting L-gulonolactone to vitamin C. Enzymes tend to get named after what they do, and in this case the enzyme is L-gulonolactone oxidase—otherwise known (fortunately) as just GLO. We thus talk of the protein GLO and the gene *GLO*. *We italicize gene names.*

As with many of our efforts to convert one compound to another, this all takes energy, not least to make GLO proteins. So what would you expect if a species switches its diet to one rich in the necessary end product, in this case vitamin C? If you can take up the vitamin from the diet in high enough amounts, there could be advantages to not wasting energy on making it. At the very least, a mutation that disrupts the ability to make it will probably not be harmful to the bearer. In fruit bats with, as the name suggests, a diet rich in vitamin C, selection to preserve the genes for synthesis has disappeared and they no longer make GLO. We are like them—we can no longer make vitamin C; we just eat it.

Also like fruit bats, in the not-so-recent evolutionary past, our ancestors too could synthesize vitamin C. There are two ways we know this. The first is that we evolved from other mammals, and nearly all other mammals (indeed most other species) can synthesize ascorbic acid. Given the across-species distribution of vitamin C synthesis in mammals and elsewhere, it is much more likely that somewhere in our evolutionary past we lost the ability rather than that everything else independently gained it.

To see this logic a bit more, think about a tree with multiple branches. The branches high up produce flowers. Imagine that on all these branches the flowers are red. Now take a cutting from a branch at the base of the tree (one not producing a flower presently) and grow that up into a full tree. What color would the flowers be? You would be right if you thought that they would be red. You are making an inference about the genetic state of cells at the base of the tree, given what all the branches are like at the top of the tree.

Now imagine that the tree has red flowers on all branches except one. Somewhere at the top of the tree is a branch that produces white flowers. Let's take that cutting again from the base of the tree. What color flowers would this make? You would still have a very good bet that it would be red. It is much more likely that the base of the tree had instructions for red flowers, but along the branch with the white flower a mutation occurred that causes flowers to be white. The alternative would be—given the structure of the tree—that the lowest branches code for white flowers but that many times independently this has mutated to red. The argument is based on the relative likelihoods of different past histories.

So too when we try to reconstruct what our ancestor species looked like. The past history of species is very much like a branching tree (we even refer to "phylogenetic trees"). Knowing the endpoints of the branches (the flower color on each branch), we can then work out at each branching event the most likely condition. We call this *ancestral state reconstruction*.

Fruit bats, guinea pigs, and some of our closely related primate species are the odd exceptions in needing Vitamin C in the diet, as they too cannot make it. Thus, as we go back through our species-level family history, we will come to a position where a now-extinct ancestor (a former fellow primate) and all its ancestors could almost certainly synthesize ascorbic acid. The best guess for when in our evolutionary past we lost the ability to make vitamin C was a common ancestor between us and lemurs.

The other reason we know our distant ancestor could synthesize vitamin C is because in our DNA is the smoking gun. We don't have a functional version of the *GLO* gene in our DNA, but we do have a dead version. In mammals with the active *GLO* (i.e., most of them), the gene has 12 exons. We and the close primate relatives have exons 4, 7, 9, 10, and 12 more or less intact; the others are either missing altogether or just a fragment of the normal full exon. Dead genes like this have a special name: we call them *pseudogenes*, meaning false genes.

As we would expect, the species with the *GLO* pseudogene have lots of vitamin C in their diets. Gorillas are estimated to consume 20–30 mg per kilogram of body weight per day; howler monkeys take in about 88 mg, and spider monkeys 106 mg. The recommended daily allowance for humans is about 1 mg per kilo per day. While in some birds we find species with high vitamin C in the diet *and* the ability to synthesize it, no species has been found with the dead *GLO* that has a diet low in vitamin C.

The evidence thus is all consistent with the notion that somewhere in our ancestry we could synthesize vitamin C but cannot now. We can also understand why a change in diet might lead to a loss of ability to make vitamin C. But notice that nowhere in the evolutionary calculation for why a lineage might lose vitamin C (or might not lose it) is there a calculation of what nineteenth-century British sailors might need (odd to say). Selection a few million years ago cannot do that. Selection cannot forward-think, "Because in several million years people (who haven't yet evolved) will want to sail across the seas and won't have enough fruit, it would be good not to lose the ability to synthesize vitamin C." I, by contrast, can go out with a raincoat because I am expecting rain even though now it is sunny. Selection can only act on mutations in the immediate here and now. There is no master plan for evolution with a pre-designated end goal.

All for One and One for All—Well, No, Not Really

If evolution's lack of foresight might seem a bit obvious in retrospect, some of its other properties are perhaps rather more subtle. If you ever want to really annoy an evolutionary geneticist, just tell them that such and such couldn't happen because it wouldn't be "good for the species," or conversely, that we do expect a certain trait to evolve because it will be good for the species. This will raise the blood pressure of the said evolutionary biologist, they will rapidly turn red, and steam will issue from both ears. You have found the secret words that really get our goat.

From the outside this probably seems like nit-picking. Evolution is sort of about being good for species, isn't it? The mutation that makes the moth better camouflaged was good for the species, wasn't it? The problem is that, even if the mutation in the end may be good for the species, it doesn't spread in the population *because* it is good for the species. It goes from rare to common because it is good for the individuals that bear it.

Put differently, in the above evolutionary model what is key is what happens when the mutation is new in the population. If the mutation cannot go from rare to a little less rare, then no matter how good it might be for the species (when everyone has it), it cannot have a future. Its down-the-line effects on the species' survival are at no point part of that calculation, just as selection cannot preserve vitamin C production because it might in the distant future be useful. Any effect it might have on the species is a consequence, not a cause of the spread of the muta-tion. And evolutionary biologists spend a lot of time trying to disen-tangle the long-term consequences from the immediate causes.

If you think this point is a bit subtle, you would not be alone. You will still find many (non-evolutionary) biologists making the same mistake, and it took quite a lot of the twentieth century for the evolutionary community to work through the logic. One of the first key problems in this context was the problem of the sex ratio. The sex ratio is the ratio of males to females. We can, for example, talk of a 1:1 ratio, meaning there are as many males as females. Look around nature and you see many cases of what look to be about 1:1 ratios—they are the norm. Why they are the norm, however, took a while to figure out, as at first sight species would be better off with sex ratios that are heavily female-biased.

Let me walk you through this logic. In species with separate males and females, the females usually do more investing in the next generation. Often this is part of the biological definition of males and females: males make small sperm, females make big eggs. Who, you should ask yourself, puts all the yolk into the kiwi bird's gigantic egg? By definition, the female. More modestly, who puts all the resources into fish eggs? Again, by definition, the female. There are other investments, such as—in mammals—transplacental resourcing and milk production. All from the mother. There are exceptions: in seahorses males do the parental care, and in many species both males and females feed the young. But generally, a good rule is that females do the great majority (if not all) of the investing in the next generation.

Now let's ask ourselves: What does this investment buy? It buys offspring. What we also know is that in (nearly every) sexual species, every baby born has two parents, a mum and a dad. We also know that the hypothetical species we are thinking about is neither an ever-growing species nor is it an ever-shrinking species (in population size, I mean, not body size). In the latter case there is nothing to discuss, as an ever-shrinking species will go extinct. If there are 100 individuals, then 90, then 80, and the numbers keep going down, there is only one end point. Extinction. Conversely, no species can forever increase its population size. Recall our example of how quickly we'd be deluged with salmon if the thousands of eggs released by just one were all to survive. Population growth has a limit too. So we can then make a good approximation: that every reproductive event must result, on the average, in two surviving progeny just as it starts with two surviving progeny, a male and a female. That isn't just two offspring (each salmon makes thousands of eggs), but on the average only two offspring per mating survive to mate.

Now we know something else. If females do all of the investment (a good first approximation), then each female must in turn have enough resources to make two surviving offspring.

And now we are ready to see the dilemma. What trait would be best for the species? Probably one that maximizes the growth rate of the population. The same goes for any economy: governments want to maximize the number of people investing in that economy. But if females do all of the investment and can make two surviving progeny, they should then make more daughters than sons. Sons are a waste of space, as they don't invest. A species could grow its numbers much faster if only the barest minimum of males were produced. Each mating needs one male and one female, but it could be the same male mating many females. Imagine a species with 100 reproductive individuals, 99 females and 1 male. This has 99 investors. By contrast, a population with a 50:50 (i.e., 1:1) sex ratio will only have 50 investors. The population should be better off with more investors, so long as the one male can fertilize all 99 females.

Here then we have the sex ratio problem. The expectation seems pretty clear: if the calculus of evolution has something to do with what is best for the species, we could understand female-biased sex ratios. But we know most species have 1:1 sex ratios. How do we square that circle?

Here again we meet R. A. Fisher. He asked a different question. Much as we have seen that when you reach perfection—the bull's-eye on the dartboard—no "better" mutation can invade, he asked whether there could be a sex ratio that couldn't be invaded, couldn't be bettered. (Here, by the way, "invade" is just shorthand for "get a foothold and spread into a population"—going from one mutation in the first generation to a higher relative frequency in the next, and so on.) Could there be a sex ratio at which a strategy producing a different sex ratio could not invade the population?

What Fisher noticed was that what looks to be best for the group or species is not such an uninvadable strategy, usually. Consider a population that is making 99 daughters for every one son. The species is doing just fine. But, said Fisher, what if a mother has a mutation enabling her to make more sons than daughters? You see, in the 99:1 population that one male is an incredibly successful male—he is the father of all off-spring! So, let's think about a new mutation that starts out rare and makes a mother produce only sons. As a consequence, this mother has resources to make two surviving progeny, both sons. As this mutation starts out rare, the sex ratio would then be 97:3. Will this mutation in-vade? It certainly will, as these two sons will collectively be the fathers of 2/3 of all progeny (on average). The average female cannot claim anything like such domination. The mother with the mutation for a male-biased sex ratio thus has loads of grandkids, and the mutation thus will be at a higher frequency down the road than the current strategy of making 99 daughters to one son.

Here then was Fisher's first big insight: the strategy that maximizes something you might want to think of as species or group fitness is an invadable strategy. Species fitness cannot therefore be the great arbiter of what we see and don't see as the product of evolution.

We can play the same thought game should this new mutation start to become common. The population sex ratio could now flip to 99 males to one female. A rare female now making a female-biased sex ratio will leave more grandkids—her female-biasing mutation will invade. So neither male- nor female-biased sex ratios are uninvadable. Is there an uninvadable solution? Under broad conditions, Fisher argued, there was one: a 1:1 sex ratio.

Actually, the argument was a bit more subtle. What Fisher argued was that the stable (uninvadable) sex ratio involved equal *investment* in both sexes. If to make a son, for example, costs more in resources, then the stable head count of individuals would not be 1:1 but would be female-biased. This insight opened the way to some downstream tests. In some species of wasps, for example, sons and daughters cost different amounts, but in other, related species they cost about the same (you can get at this by asking about their weight minus their water content, so-called dry weight). Rather nicely, in all species the unifying variable was that the net investment in males and females was equal in all the species examined.

The framework outlined by Fisher really helped clarify thinking on how to predict the outcomes of the evolutionary process, and why the key question is what happens when a mutation first appears in a population, not whether it is "good for the species." That is, are the circumstances surrounding its first appearance "good for the mutation"? This isn't the complete story of sex ratio evolution, however. As we will see, this account has been simplified by making a number of assumptions that are often, but not always, true. Importantly, the consequences of exceptions to these assumptions can be understood by asking whether the new mutation can or cannot invade. For example, Fisher assumes that the mutation controlling the sex ratio is not on one of the sex chromosomes. This is a fair assumption, as most genes are not on sex chromosomes. A gene on a Y chromosome that controls the sex ratio, for example, would be uniquely favored to make more sons (since only fathers and sons have Y chromosomes). And indeed, we see odd male-biased sex ratios in some mosquitoes whose gene controlling the sex ratio is on the Y chromosome.

Fisher also assumes that any male can mate with any female. To receive the strong "2/3 of matings" advantage, we assume all males are potential fathers to all offspring. In some wasp species this isn't the case. These aren't the wasps that make a misery of summer picnics, but a different sort. In these cases, a mother injects a full family of ready-to-develop single fertilized egg cells inside the body of one caterpillar. These develop into her sons and daughters, and these sons and daughters mate within the caterpillar (yes, incest). The pregnant females then escape and go to find a different caterpillar to inject.

In this case, a mother making more sons will not be at an advantage, as these sons can only mate with the other residents of the same caterpillar. It is a bit like being put in a room for nearly all your life and told that you can only mate with other residents of the same room (your siblings). In this case, the "good for the species" calculation and the "good for the mutation" calculation agree that the uninvadable strategy is a heavily female-biased sex ratio: make only as many males as needed to fertilize all females. And this is what we normally see.

I say "normally" because the "good for the mutation" model makes another prediction. Imagine a female puts her young into the caterpillar with a female-biased sex ratio—just enough males to fertilize all the females. Now imagine a second female comes along and puts her off-spring into the same caterpillar. This would be like opening the door to the room you are in and another family being introduced. If the second mother wasp knows that she is not the first to place her family in the caterpillar (and somehow often she does), she then tweaks her sex ratio. The British evolutionary biologist W. D. Hamilton (commonly known as Bill) calculated what would be this second female's best sex ratio to produce. He determined that it should also be female-biased—but not quite as female-biased as the first female's sex ratio. The reason for this is that the second mother's sons can also mate with the females of the first brood, so making more sons gives a bigger contribution to the next generation, genetically speaking. Experiments in wasps find that Hamilton's model predicted the second female's sex ratio quite well. And the third, fourth, etc. The "good for the mutation" calculation gets it right again.

Can I Have More, Please?

The view that the logic of evolution is not about what is good for the species has many odd ramifications. Perhaps some of the oddest are when we think about mammals and mother–fetus interactions. Surely, here is a case where what is best for the species is for mothers to look after their fetuses and fetuses to look after the mother? Surely, selection would operate to promote the best for both? As Bob Trivers of Rutgers University was the first to notice, this isn't necessarily so.

He pointed out that a mother's "best interests" are in looking after all her kids equally, as she is equally related to them all. They are all equally likely to have any new mutation that has happened in her. The same need not be true for the offspring, especially if they don't all have the same father. Full sibs (with the same father) are more likely to share mutations than half-sibs that only have a mother in common. Even if there is no change in paternity, any given embryo is more related to itself than to its sib mates, but the logic is easier to see when there is a change in paternity. Incidentally, in talking about "best interests," we are actually using a shorthand for mutation invasion (or non-invasion).

One particular manifestation of this problem comes in species like mammals where there are discrete broods. In this instance the father of today's offspring might not be the father of a subsequent brood. In this case the mother is related equally to all her kids, but today's kid will not be as related to their half-sibs in the future (they don't have the same father). As Trivers pointed out, the mother's "best interests" are to hold resources back for her future kids. By contrast, a fetus can commonly be better off getting more resources from the mother now—the future half-sibs they are hurting by doing so don't necessarily have the same mutation affecting resource demands, as they don't have the same father.

This would be a case where we have a so-called "conflict of interest": what is best for one party is not what is best for the other, and vice versa. Conflicts like this can play out in many ways, and predicting where evolution will end up is tricky. When there are conflicts many different outcomes can be gamed, which makes such predictions somewhat open-ended. That doesn't mean that conflicts don't happen, it just makes such ideas harder to test.

With this caveat, the conflict model does seem to make sense of some odd data. David Haig of Harvard University has, for example, suggested that such tensions may explain why human fetuses put out chemicals into their mother's blood system that keep sugar in the blood, enabling more resources for today's baby in the womb, potentially less for one further down the line. For example, the placenta (a fetal tissue) releases human placental lactogen, which has anti-insulin properties, into the mother. This stops glucose in the mother's blood from being removed to the mother's cells for longer-term storage and keeps it available for the fetus. In turn, mothers produce more of the cells that make insulin (beta cells in the pancreas), thus shifting the balance back toward the mother keeping hold of the precious sugar for longer-term use. If Haig is right, this isn't two parties agreeing on the best blood sugar level, but more like two parties pulling the two ends of rope in a tug-of-war. Complications resulting from this delicate balance can, it is suggested, lead to complications in pregnancy such as gestational diabetes (too much sugar in the mother's blood).

Similar tug-of-war games are thought to possibly control maternal blood pressure. Higher pressure means more food for the fetus, especially early in pregnancy. Indeed, studies of many thousands of babies in Denmark find that babies born to mothers that had higher blood pressure early in pregnancy tend to be larger and are more likely to survive. In late pregnancy, human mothers are especially prone to high blood pressure, which may be part of this tug-of-war.

However, general high maternal blood pressure, which may be good for the baby, needs to be distinguished from the more complex condition of pregnancy called pre-eclampsia. This is associated with very high maternal blood pressure and protein in the mother's urine. Both features are used for diagnosis. This seems to be both common (5% of pregnancies) and human-specific, and it is deadly. The only cure is rapid delivery, for example by emergency Cesarean section, to whip the baby out before the mother's blood pressure goes off the scale.

It looks to be a consequence of a stressed placenta being attacked by the mother's immune system, and possibly responding by increasing blood pressure to increase resources to the placenta and thus the baby. That maternal immune rejection of "unknown" proteins from the fetus might be the underlying cause is suggested by the fact that it can also be prevented by having lots of unprotected sex prior to getting pregnant. The idea is that the man's semen somehow trains the mother's immune system to recognize his proteins, and not therefore immunologically reject the fetus when the mother's immune system sees it. Indeed, if we look at placentas of pre-eclamptic mothers we often see a protein, "sperm-associated antigen 4" (SPAG4) overexpressed. This protein, expressed in sperm, is also found at that maternal–fetal interface. That mothers carrying other people's babies (surrogate mothers) are the most likely to get pre-eclampsia fits with such an immune model.

As pre-eclampsia can easily end up with the death of the mother and fetus, it doesn't sit so comfortably in the tug-of-war framework. It is good for neither mum nor baby and seems to be a runaway process. I have a suspicion that, as it affects so many pregnancies, it must be one of the most important causes of mortality (and hence selection) in human history. Why it is so common remains an unexplained enigma.

The Hidden Assumptions

To understand evolution, then, we must think about new mutations and their fate. But beyond this, and the fact that mutations are just changes to DNA, where is the genetics? As I hinted, there were several (rather hidden) genetic assumptions in the simple model of selection that I presented. For example, I assumed that when a new mutation arrives in a population, its effects will be visible. This need not be the case. In many species, including the moths we were thinking about, individuals have two copies of every gene. When the mutation first appears in a population the individual will have one version of the gene that is the old version and one that is the new one. But why should such an individual be a little mottled? Could not the mottling effect be masked by the other, older, version of the gene?

In technical language, I assumed that the new mutation is at least "semi-dominant"—it has some effect on what the moths look like when partnered with the old version of the gene. A fully dominant new mutation is one where individuals with one old, one new version look like individuals with two new versions. Unfortunately, some disease-causing mutations are also dominant—for Huntington's disease, for example, you need only one bad version of the Huntington's gene to get the full-blown disease. The new mutation could, however, have been *recessive*: its effects could have been fully hidden. The mutation that causes cystic fibrosis is like this—you need two of a bad version to get the condition.

What if having both copies of the new version renders you mottled, but having one old and one new makes you look like the old gray-black? The British geneticist J.B.S. Haldane noted that selection will impose another sieve: mutations that are beneficial *and* genetically dominant are more likely to do well when rare, and so will be more likely to contribute to adaptive change, because what happens when mutations are rare is so important. This is known as "Haldane's sieve"—it is a dominance filter on the filter that is positive selection. There are, however, even more complexities to dominance that further complicate our understanding of evolution. What, for example, if the very best type has one of the resident version of the gene and one new mutation?

Another key assumption concerns the mutation rate. I emphasized how selection on one gene causes it to go from rare to common much faster than new mutations come in to disturb the process. What would happen if the mutation rate were really high? Can we still have survival of the fittest if the fittest are degraded by mutation really fast? For the most part this isn't an issue, as mutation rates tend to be rather low. But it may be important for some viruses.

There are two other crucial assumptions that I very quickly skipped over. The first of these is that transmission is unbiased. Note that I assumed that an individual with one old, one new version of the gene will transmit the new one to exactly half of the offspring and the old one to the other half. This isn't such an odd assumption, biologically speaking. The process of making sex cells (sperm and eggs) is strangely complicated, but what it does normally results in unbiased transfer of mutations from parents to offspring.

When we make each sex cell, we go from a cell with two copies of each chromosome $(2 \times 23$ in our case) to cells with one copy of each chromosome (1×23). This isn't 23 randomly selected chromosomes, but one of each—i.e., one chromosome 1, one chromosome 2, etc. Sperm with 23 chromosomes then fuse with eggs with 23 to make a new human with $2 \times 23 = 46$.

The way we do this usually involves a sort of fair coin-toss-like process. In males, for example, each cell with two copies of each chromosome makes a cell with four of each (4×23), which divides to make two cells with two of each (two lots of 2×23), each of which divides to make two more cells, each with just one of each chromosome (4 cells with 23 each). I told you it was strangely complicated.

Key to the process is that at each cell division, each chromosome pair lines up—they behave like dance partners: one copy of chromosome 1 pairs up with the other chromosome 1, chromosome 2 pairs up with chromosome 2, etc. Each cell division then pulls apart one copy of each chromosome to each daughter cell. Thus, at each division, if the chromosome with the new mutation turns left, the chromosome with the old copy of the gene goes right. Thus, exactly half of the cells have the new version and half have the old version. Unbiased, 50:50 segregation is thus the default state. But what if the transmission is somehow biased? Will the fitness effects of the new mutation be the only thing that matters, or will the transmission rates also play a part?

My second—very well-hidden—assumption is that the only thing that matters is whether a new mutation is good or bad for you. I ignored chance. But chance can be important. Imagine when the mutation first arrives there is one moth that has both this new mutation and the old version. It stands a better chance of surviving because of what it looks like (because of its genes). But what about when the moth's sex cells get to randomly meet another moth's (with only the old version)? Imagine the mating results in two offspring. Half the time, just because of chance, one offspring will have the new mutation and one will not. A quarter of the time, just because of the luck involved in which sex cells get to fuse, both offspring will have the new mutation. A quarter of the time neither will have the new mutation. On the average, the mutation neither goes up nor down in frequency (50:50 transmission as we saw above)—but when the mutation is new in the population, its effects on fitness can be as nothing compared to sheer bad luck. Remember, a quarter of the time this brilliant new mutation will be lost simply because, owing to chance sperm-egg encounters, the sex cells with the new mutation don't get to be in either of the offspring.

For all mutations, the time of greatest jeopardy is indeed when they are rare. They can very easily be lost from populations because of chance alone. Indeed, J.B.S. Haldane calculated that the chance of a new mutation that increases fitness by some percentage—let's say $x\%$—manages to actually get to fixation is about 2 times x. So if a mutation increases your fitness by 1%, say (a big effect), it has only a 2% chance of getting to be fixed. The other 98% of the times it is lost, because when you are rare you also need to be lucky to get to be common enough not to be lost by chance alone.

As we shall see, if any of these assumptions are broken, evolution need not (and commonly does not) lead to perfection. The last two—chance sampling and unbiased transmission—seem to explain many of our genomic mysteries. Let's start by taking a deeper look at the problem of chance sampling. It has some odd properties.

4

The Great Evolutionary
Roulette Wheel

In 1943 the US Air Force was running daylight bombing raids over Germany and suffering substantial losses. The military decided that something must be done. They could add armor to planes, but this is heavy and would lower a plane's performance. What, then, is the optimal amount of armor to add, and where should it be added?

To answer this question, they called on the services of the US Statistical Research Group (SRG), a then recently formed and highly talented group. It included many individuals who, postwar, would rise to fame for other reasons. There were two economics Nobel Prize winners, Milton Friedman and George Stigler. By legend, it was, however, "the smartest man in the room," a Hungarian Jewish refugee, Abraham Wald, who was tasked with solving the plane reinforcement optimization problem and the statistics of plane losses more generally.

What Wald first needed to know was where aircraft were damaged. The military had done a survey of bullet holes on its planes, and this was Wald's raw data. It had been noticed by the military that there was often damage to the ends of the wings and the fuselage, but relatively little around the engines or cockpit. According to an oft-told story, the military were thus expecting to be told exactly where on the wings and fuselage to judiciously add the armor. But Wald advised the very opposite.

The planes that were examined, he reasoned, were the successful returning planes. Where they had damage showed where the plane could be hit and still fly. This is known more generally as "survivorship bias." It is from where the surviving planes were *not* damaged that you could infer where the planes were the most vulnerable. Hits to the cockpit and engines were the ones that brought planes down. And, so the legend goes, the planes were correctly reinforced, and many lives were saved.

On the internet you will find many incendiary stories relating this history along the lines of "Mathematical genius scores against military," but actual evidence for this legend is rather thin on the ground. Indeed, what of the above story is true and what invented is hard to work out (and key papers are buried in US Navy Archives). It seems unlikely that any specific advice was given, as SRG usually just answered the questions asked, rather than making policy decisions—that was the military's job.

What seems to be clear is that most of Wald's work was related to a different problem: working out how many hits a plane could take and still return safely. But from the notes of a contemporary (W. Allen Wallis), we know he did also work on statistical methods to estimate the vulnerability of various parts of planes to bullet holes. There is some kernel of truth to the legend. However, beware of internet accounts. Many feature images of idealized planes with an excess of bullet holes at wing tips and fuselage—planes that never flew bombing missions over Germany.

Legend or not, hidden in the two problems that Wald examined—how many bullets a plane can take and where on a plane parts are more vulnerable—lie truths about the nature of chance and selection. They apply every bit as much to the evolution of organisms and their DNA as to World War II bombers. Mutations are like bullets. The metaphor isn't quite perfect, but let's leave that to one side for now, as the principles are similar.

One evident truth is that some mortality has a chance component. Two planes flying through the same anti-aircraft fire will, on average, have the same chance of returning. But one may be unlucky and get more bullets, or bullets in more vulnerable places, just by chance. This need not be because of bad captaincy or anything like that. So, too, some individuals will die owing to bad luck, as they inherited more mutations than others, or the mutations were in the wrong place. Indeed, while we have many environmental predictors of who gets cancers (a more direct analogy of bullets hitting the plane), a major component seems to just be bad mutational luck. In these instances, these are mutations in our bodies that happen as we get older.

Chance affects not only how many mutations/bullets hit, but also where they hit. The idea that some parts are more vulnerable than others (engine and cockpit are vulnerable, fuselage and wing-end less so) applies also to DNA. Indeed, mutations hit DNA randomly, but their effects are very different. A mutation that stops a key protein from working will likely kill you. A mutation in a gene that converts one codon to another giving the same amino acid—e.g., GGC → GGT, both coding for the amino acid glycine—may well be less likely to have a big effect. A mutation in a pseudogene (remember dead genes?) could well have no effect on fitness, which explains why they decay over time after becoming dead genes. DNA, then, is like an airplane in that bullets to some bits are likely to be fatal, but the same bullet hitting elsewhere is less likely to have an effect. There is variation in the fitness effects of mutations.

In this chapter we will see how, as a species, we are especially prone to accumulate, by chance, mutations that don't affect us very much. Over time we end up riddled with bullet holes at the wing tips and in the fuselage. Why we are especially prone to such mutation accumulation hinges on a subtle difference between tossing coins that have a 50:50 chance of being heads versus tails or betting on a roulette wheel, where your chance of winning is ever so slightly less than 50:50. As we will see in the following chapters, it turns out that this provides our best explanation for many of our imperfections: why our DNA is bloated, functions badly, and mutates a lot. To start with, let's look at the underlying theory.

Using DNA Sequences to
Understand How DNA Evolves

First let's consider how, using insights related to Wald's, we have a good idea about how DNA evolves. We have a good guide as to which bits of our DNA are more or less vulnerable, and it is logically the same as

Wald's insight. Imagine matching up the anatomy of the returning planes against what they looked like before heading over Germany. The sites that are *different* in the surviving planes tell you where sites are less vulnerable: the planes have holes at the ends of the wings that they didn't have before setting out. You never get to see holes in the engine in the returning planes because the planes with holes in the engines went down. The apparently unchanged (conserved) bits in the survivors are the important bits.

So too our genes. As we are all evolutionary survivors, if a site in our DNA is crucial, we expect that its current state is the ancestral condition, just because those that were hit never returned to base. Remember, we can do ancestral state reconstruction if we know the genes from related species. Indeed, we find that parts of genes that code for the key parts of proteins evolve more slowly than other parts of the proteins, which in turn evolve more slowly than sites in codons that don't affect the protein produced. Generally, slow evolution is a great guide to functional importance.

We use this insight routinely in trying to work out which of the many mutations you inherited might be the one causing a genetic disease. Sites that are slower-evolving (the site is the same in us, chimps, gorillas, etc.) are likely to be functionally more important: the engines and cockpits encoded by our DNA. And let's be clear why this trick works: it isn't because the slow-evolving bits are hit by mutations any less than the faster-evolving parts, just as different parts of planes are no more or less likely to be hit by bullets. Rather, it is because the mutations/bullets that hit different parts have different effects and different chances of being in a surviving body/plane.

There are, however, complexities in the application of this insight, as the rate of mutations varies by genomic context. For example, some parts of our DNA do have slightly higher mutation rates than others. This would be a bit like different parts of planes being more or less likely to be hit by bullets—for example, if those shooting tended to aim toward the parts without bulletproofing. The insight that purifying selection leads to slow evolution (a mutation appears in the population but disappears as the body dies; a bullet hits a plane, but the plane never returns to base) also gives us a routinely used fingerprint of purifying selection acting on genes that code for proteins. With some tricks, we can control for mutation rate differences.

Commonly, what we do is to compare rates of evolution at sites in protein-coding genes that would change the protein with those that would not in the very same gene. Remember that we have 61 codons that specify an amino acid but only 20 such amino acids. Consequently, many changes to the DNA swap between codons that code for the same amino acid. GGC and GGT both code for glycine, for example. We use the rate of evolution of the "synonymous" (synonyms mean the same thing) codons as a measure of the local background rate of evolution. That way, any differences in mutation rates around our DNA are taken out of the equation (or so we hope). We can then compare this background rate to the protein rate, and by so doing, find the sections of genes that code for the important bits (the engine and cockpits). This is known as the *Ka/Ks ratio test*. You will also see the same test sometimes called dN/dS and the ratio referred to as "omega." I once again need to apologize for the confusing terminology in my field. Let's just call the metric Ka/Ks, as the different metrics are at heart all the same.

To calculate Ka/Ks, we need the same DNA sequence from at least two different species. Well, that isn't quite true—you can also do it using at least two sequences from the same species, but the test is weaker as there hasn't been a lot of time for changes to happen and be acted on. This is a bit like sampling planes before the end of the bombing mission. Indeed, if you sample them just as they took off you will see no differences between various parts, as neither mutations (bullets) nor selection (planes not returning home) will have happened.

You start by lining up the DNA sequences so that each codon in the gene of our first species is matched to the same position in the other species. This gives us a "sequence alignment." The same gene in different species we call *orthologs* and the same codons (by position) are then also orthologous codons. For distantly related species, alignment can be a headache and source of error (the correct orthologous codons don't line up), but for most comparisons within, say, mammals, alignments of protein-coding genes are usually pretty tidy.

From this alignment, we count the number of differences that have changed the protein and the number that didn't. To see this, let's just consider a small section of a protein-coding gene. Imagine in us (it doesn't have to be us, but run with it) a gene starts: ATG GGT GCA TTC. This codes for 4 amino acids: ATG = methionine, GGT = glycine, GCA = alanine, TTC = phenylalanine. Now let's look at the very same segment, but now in a different species—mouse, perhaps. Imagine in mice the sequence runs ATG GGC GTA TTC. The first amino acid and first codon are the same: ATG. No evolution. The second has changed: GGC is seen in mouse, GGT in human. With just two sequences we don't know whether GGT changed to GGC or vice versa (or in principle there could be more complex possibilities). What we do know is that the codons are not the same. Evolution. But in this case, GGC and GGT both give the same amino acid, glycine. The DNA sequence has evolved but the protein has not. We add one to the count of "synonymous" changes. In mice the next codon is also just one mutation difference from the human sequence GTA versus GCA. But GTA codes for the amino acid valine while GCA specifies alanine. Both the gene and protein have changed. We add to the count of protein-changing mutations. The last codon is, like the first, the same. No evolution, we don't count anything.

We can get our computers to scan all our genes like this, aligned with not just mouse but any number of species. The more species, the finer the resolution of which sections of our genes code for engines and which code for wing tips. The "engines" will have especially low rates of amino acid swaps compared with the background rate.

Before we can go there, however, we have some statistical corrections to make. It is, for example, the nature of our genetic code that random mutations will change amino acids about three-quarters of the time, so we need to correct for this difference in the number of synonymous and nonsynonymous sites. There might also have been multiple changes at a site, but we see only the end states (two bullets hitting the same place in the plane becomes more likely if many bullets have hit the plane). We can correct for that by knowing how much change we see overall—the more changes we see, the greater the number of unseen changes.

Finally, we come up with a number—the ratio of the number of non-synonymous changes per nonsynonymous site (Ka) to the number of synonymous changes per synonymous site (Ks). The former is a measure of the rate at which the protein evolves, the latter the background rate.

In the field of molecular evolution the ratio of these two, Ka/Ks, is a magic number and does something Wald could only dream of. He never really knew the number of bullets that hit the planes that went down. We, on the other hand, can ask: How fast is this protein evolving compared with the background rate of evolution?

If a protein is subject to purifying selection, we expect Ka/Ks to be less than one. That is to say, allowing for how many bullets have hit the plane (Ks), there are fewer bullet holes than would be expected if no bullets damage the plane's chance of returning to base (Ka). A ratio of Ka/Ks less than one thus tells you that the protein is subject to purifying selection, just as planes that were hit in important parts by bullets never return to base. If we do this comparing, for example, mouse genes with their human equivalents we see that Ka/Ks is usually much less than one, usually about 0.1 (fig. 4.1). For every ten mutations that changed the protein, only one repeatably made it back to base (the organism and its descendants survived).

This confirms an obvious but profound prediction of the dartboard model we saw before: because we are a reasonably well-functioning organism, most mutations that hit functional sections of our DNA are harmful, not beneficial. Similarly, if you make random tweaks to a Swiss watch you are more likely to make it worse than improve it, and bullets tend to make planes worse, not better. This is a general feature of any structure that, for whatever reason, has a relationship between its form (engine, protein shape, watch escapement) and function (flying, living, telling the time). Most mutations, we assume, are thus either selectively irrelevant (if in non-functional DNA) or harmful.

Some genes are so intolerant of mutations that change the protein that they have Ka/Ks equal to zero. This means that no change to the protein can be tolerated—i.e., very strong purifying selection. These are like planes that go down wherever on the plane they are hit: our most vulnerable genes. Interestingly, our proteins associated with genetic

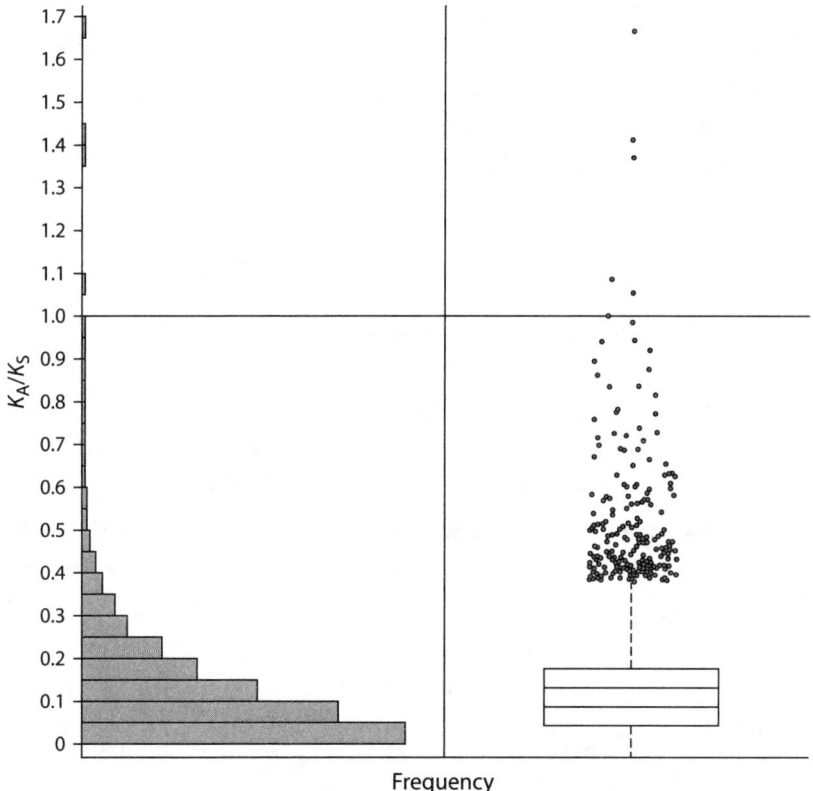

FIGURE 4.1. **The distribution of Ka/Ks values for 5,057 mouse-human genes.** The distribution of values is shown in two ways: as a histogram on the left and as a box plot on the right. Adapted from: Hurst, L. D. 2009. "Genetics and the understanding of selection." *Nature Reviews Genetics* 10:83–93.

diseases tend not to be these especially slowly evolving genes. At first sight this appears paradoxical. Surely genes that are most vulnerable to mutations would be the ones most likely to cause genetic diseases. But it isn't so. A genetic disease is defined in terms of humans that have survived at least to birth. If a gene and all its parts are so vital that mutations kill you as a very early embryo, then we never get to see these, and they don't get classified as disease-causing mutations. Our disease-causing proteins instead tend to be medium-slow evolving: they are important enough to be under purifying selection, but not so important as to kill you when you are an embryo in the mother. Interestingly, they also tend to be longer than the average protein. This length effect is also owing to an "ascertainment bias": long proteins are like bigger planes—they are more likely to be hit just because they are bigger, so they come to our attention more, as they cause more disease.

Some Genes are under Selection to Change

We also sometimes see genes that have Ka/Ks greater than one. This is not something the US military had an issue with. What this usually means is that the changes to the protein increase the chances of surviving—these are beneficial mutations. Planes hit by bullets are not expected to increase their chances of returning to base.

Looking across species, we see regularities in which genes these are. Proteins associated with big organisms interacting with their parasites (meaning any species that infects and harms us, be it viral, bacterial, multicellular) and proteins in the parasites that interact with the hosts' immune defense systems are one such category. Often genes associated with male–female interactions or mother–baby interactions are also ultra-fast evolving.

In all these cases we suspect an underlying similar rationale: a conflict of interest. What is successful for a parasite—like the malaria parasite or the viruses that cause dengue or yellow fever—is not what is best for us. Selection on them is to use our bodies as their incubator and be transmitted to another host. The ones that are better at this are the ones we will see more of. Conversely, over the longer term selection favors those hosts that are better at not being used as a parasite incubator, or at least not being harmed by being an incubator. There is an analogy with Wald's plane problem: the need to reinforce the planes is a similar sort of co-evolution with the adversary firing bullets. But every beneficial change in the host that spreads in the population just changes the context for the parasites. And so, we play an eternal game of evolutionary cat and mouse with our parasites. When we look for DNA-based signals of positive selection across human populations, it is striking how many of the signals are in genes that mediate our interactions with mosquito-transmitted parasites (like malaria, yellow fever, Lassa fever, etc.). Human evolution would look very different were it not for mosquitoes.

The same cat-and-mouse game is likely to be true for male–female relations and mother–baby relations. As we saw before, what is best for the baby—give me more resources—is not what is best for the mother—hold some back for the future. And so here too we expect an evolutionary game of cat and mouse. Many of the proteins involved in these interactions, such as human placental lactogen, are indeed fast-evolving.

Males and females are also thought to be engaged in eternal conflicts. Consider, for example, fruit flies. After mating, females undergo many changes: they become less attractive to males and less receptive to males; they increase their egg-laying rate and utilization of sperm. They also have a decreased life span. While some of these changes may benefit the female, they all seem to benefit the would-be father. We also know how these changes come about: they are mostly caused by a variety of substances transferred along with the sperm into the females on mating. They are the products of a male's accessory reproductive glands and are accordingly called Acps (accessory gland proteins). Many are some of the fly's fastest-evolving proteins.

There is a rich diversity of these—an individual male makes 80–100 different sorts, and it is the lifetime's work of many researchers to try to work out what each of these is doing. Acp26Aa increases the egg-laying rate, while Acp70A stimulates longer-term production of eggs. Acp36DE enables the uptake and storage of sperm. It is unusual in that it goes to the wall of the female's oviduct, close to the opening of the female's sperm storage organ (unlike humans, flies store sperm a bit like hamsters fill their cheeks with seeds, for later use). It seems to corral sperm into this strategic position.

While we can interpret much of this as manipulations by the male that are good for the male, are they necessarily part of a cat-and-mouse game? Are they also harmful for females? Do females, in turn, counter-evolve? The reduced longevity associated with mating would suggest that what is good for a male is bad for a female. Do then females evolve to suppress the toxicity of seminal chemicals?

One of the best demonstrations of this came from Bill Rice of the University of California, Santa Barbara. He reasoned that if males and females are playing long-term games of evolutionary cat and mouse, then if he could stop the females from co-evolving, the males should evolve to be even more damaging. To achieve this, he held a stock of virgin females that were not evolving and used these as mothers in each generation. Their daughters didn't become mothers in the experiment, so the female lines couldn't counter-evolve. The sons were retained in the experiment. The males could evolve, the females couldn't. Clever. The males' fitness went up over time compared with controls. Importantly, the evolutionarily static females, acting as partners to these males, did worse. The males tended to mate more, but the females, after laying the eggs, were more prone to dying.

The Problem of Mutations of Small Effect

With tests like Ka/Ks we can identify the regions of our genes that code for the plane's engines and cockpits, the bits of our DNA so important that their evolution is slow. In these parts of our genes we see fewer differences between species than we would have expected given the number of bullets sent their way. These are parts of our genes whose evolution is in some ways very predictable: if a bullet takes out your only engine, you will not be returning to base. But what about the bullets that don't have much, if any, effect? How common are these? How do the affected parts of our DNA evolve, if at all?

As we will see, recent advances in the study of DNA suggest that mutations that very slightly reduce your chance of returning to base are key to our evolution: not only are they very common, but we can't purge them while many other species can. The reason we appear to tolerate them is because we have a relatively small population size, and species with fewer individuals are more at the whim of chance events. As a consequence, our DNA degrades over time, and our cells resemble a chaotic city full of cars not stopping on red, pedestrians routinely getting run down, and buses going to the wrong place. This is the new view of what it is to be human.

To see this, let's return to Ka/Ks. As we saw, the way DNA evolves tells us that most proteins evolve under purifying selection (Ka/Ks <1: some bullets bring the plane down), but a few are under positive selection (Ka/Ks >1: some random changes make for a better plane, especially when in an evolutionary cat-and-mouse game).

But there is one final possibility: Ka/Ks = 1.

What could that mean? For a protein as a whole, it means something a bit ridiculous: that no amino acid in the protein actually matters. We do sometimes see proteins that, when compared to just one other, have Ka/Ks = 1, but when we look at the same protein across many species, we see a different story. If we have data from many species then we can ask about Ka/Ks for each codon, not just for the protein as whole. What we then see is that within a protein, some codons are under positive selection (Ka/Ks >1) and some under purifying selection (Ka/Ks <1). When averaged across the gene this can give Ka/Ks of about 1. These also tend to be cat-and-mouse genes.

We must still ask the more general question: just as we saw that bullets in planes have differing effects on the fate of the plane, are there bullets that have no or little effect on the plane? Are there locations of bullet holes where the plane is as likely to return home with or without the bullet hole? Are there mutations in DNA—not just in the protein-coding sections—that have no or very little effect on fitness, neither especially beneficial nor especially detrimental? Perhaps in many proteins we can mutationally swap one amino acid for another with no consequences? If so, how will these sites evolve? What about mutations in introns and in non-coding sequences between genes?

At first sight this looks like *one* question about mutations with no or little effect. In evolutionary terms, it turns out that it makes a considerable difference whether we ask about mutations that have *no* effect or those that have *very little* effect on fitness. The evolutionary fate of mutations that have little effect is now thought to be key to the understanding of trends across species in matters like the genome bloating that we saw in chapter 2 (number and size of introns, amount of between-gene DNA, etc.) as well as the mutation rate.

What is crucially different about mutations that have no effect and those that have some but little effect is that the rate of evolution of the former is not expected to vary with population size, just with the mutation rate; while when we factor in the latter class, the rate of DNA evolutionary change is higher when populations are smaller (like ours was until recently). This needs unpacking, not least because there is a lot of confusion about how chance influences evolution in small and large populations.

Why Population Size Doesn't Matter
if Evolution Only Depends on Coin Tosses

If we return to the 1960s, while the world was captivated by peace, love, and flower power, the evolutionary world had its mind on other things. Two of these were foremost. One was why, when we looked inside populations, we saw a lot of "polymorphism." Polymorphism is the presence of more than one form of the same bit of DNA or protein in the same population. One way we can measure this is by asking how often your two versions of the same gene (one from mum, one from dad) are themselves different at the DNA sequence level. Perhaps there are two different versions of the same gene for an enzyme with perhaps only one amino acid difference between them. The classical selection-based theory said that, for the most part, for any given gene there should be a best version (cat-and-mouse genes are different here), hence all individuals in a population should have the same best version, and thus, your two versions of the same gene will be the same. But we don't see this. The earliest studies looked at flies and found that for about 40% of their proteins, a population of flies had two different versions. This didn't make sense if there should be one best version.

The second problem was how fast genes were evolving. The Japanese geneticist Motoo Kimura took calculations from J.B.S. Haldane about how fast genes could evolve owing to selection and found that the rates of evolution of different proteins seemed to be too high. There should be brakes on the rate of evolution imposed by the amount of death a species could tolerate. If you think about white and black moths, for example, the change in frequency of the two forms is owing to greater death rates of one type depending on the background available to rest on. But Haldane asked, what if you have many genes under selection at the same time? There must be limits on the amount of change, as all organisms cannot die. The brakes that Haldane calculated put an upper limit on how fast proteins could evolve. But, Kimura calculated, they seem to be evolving much faster than this.

Kimura then argued that there is a simple solution to these two paradoxes. This is that some mutations have no effects on fitness and just change frequency by chance. In his language, there was a whole load of mutations that were "neutral"—not good for you, not bad for you, just no effect.

The evolution of neutral mutations was not owing to processes like planes being shot out of the sky. It was more like gambler's luck: chance.

If the mutations that make for differences between our DNA and that of chimps, say, are neutral, then the excess death problem goes away—there is no death associated with this sort of evolution. The polymorphism problem could also disappear. These mutations with no effect on fitness go up and down in frequency like particles of dust floating about in the room. They may eventually come to rest on the ceiling or the floor, but they spend a lot of time just going randomly up and down in frequency. Sample a room at any time and you will find dust particles floating about before they land somewhere. So too, Kimura suggested, mutations with no effects will bobble about randomly, going up and down in frequency owing to chance alone. Sample a population at any time and you will find different versions of the same gene because the differences have no effect.

We already saw part of the role of this sort of chance. When a mutation first appears, even if beneficial, chance sampling when it is rare means it stands only a low chance of getting from rare to common ($2 \times$ the percentage benefit). A mutation that gets to be the only flavor of DNA at a given position in a population we say to be "fixed" or to have gone to "fixation." However, as Kimura pointed out, mutations can also go from rare to common, and become fixed, by chance alone. Selection isn't the only mode of DNA change over time. This set of ideas constituted the first serious challenge to Darwinian selection as a mode of species change over time.

To see how this works, let's think about a mutation that has absolutely no effects on an organism's fitness. When it arrives by mutation it will first appear in the genome with one old, one new version. There are now many chance-based reasons that it could become more or less frequent in the population. Part of the chance comes about because of the nature of how we make sex cells. On average, half our progeny get one version of each gene, the other half gets the other: you have a 50:50 chance of heads versus tails. Let's now think about what happens with two kids. With two kids from each mating, a new mutation will, 25% of the time, be in both. It has gone up in frequency (from one to two copies) not because it is a brilliant mutation, not because the sperm or eggs cells it is in stand more of a chance of being successful. Just pure and simple luck.

This need not be the only reason for chance changes. Remember, organisms can be like unlucky bombers and just happen to also get more mutations, or those mutations might, by chance, hit key parts. A bullet hole of no effect will not get counted (get to return to base) if the same plane is also hit by a lethal bullet. We also have external fate. We are all prone to the slings and arrows of outrageous fortune. Similarly, even if a mutation does you no harm, if it first appears in a genome that also has a new lethal mutation, the innocuous mutation is going nowhere. If an innocuous mutation just by chance first appears in an individual who falls under the proverbial bus through bad luck, our innocuous mutation is similarly thrown under the bus.

Chance events will cause some mutations of no effect to go up in frequency and some to go down. What we want to know is what the rate of change between two species is expected to be because of this chance process alone. One way to crack this problem was to apply a branch of mathematics called branching theory. This was Haldane's approach, and it was how he came up with his estimate of the chance of fixation of an advantageous mutation. This sees each possible fate (up, down, no change, etc.) as branches in a process. However, Kimura saw a different mathematical route, inspired by particles of dust floating about.

The problem of different versions of the same bit of DNA with two different—but equally good—forms was the same as the problem that statistical mathematicians, notably a brilliant Russian, Andrei Kolmogorov, had thought about when trying to work out how gas particles diffuse over time. Odd to say, this is called diffusion theory. As is often the case, the same problem was solved by others independently. What physicists know as the Fokker-Planck equation, Kolmogorov independently derived. Kolmogorov added something that Kimura found very useful: the ability to work backward in time as well as forward. For clarity, let's follow one gas particle forward over time. It bobbles about, going vertically up or down (for the analogy we are only interested in the vertical extent of the particle's movement in one dimension—up, down). In the translation to mutations, up and down vertically is the same as up and down in relative frequency: more common, less common in the population. This bobbling about in genetic theory has a different name—we call it *drift*. But I prefer "bobbling about."

Because mutation is a rare process, a new mutation starts out really close to the floor of the room our gas particle is in. It has two fates: it can get stuck to the floor, or it could bobble about and eventually get stuck to the ceiling. The first case is the new version arriving by mutation and being lost because of chance, either immediately or eventually. The second is the mutation going from rare to common and becoming a fixed difference, again because of chance (the Kolmogorov backward equation allows you start at this fixation and work backward). In the latter case, if we compare this species' DNA to that of a close relative (say human–chimp), it will appear as one of the DNA differences that defines us as different from chimps.

Compare that with selection in the case of, for example, our moth. There, too, a mutation could go from rare to common, but this is not owing to chance—this is owing to the black moths surviving better on black backgrounds. The fate of the mutation is owing to the action of the mutation: making a moth black. If you were to repeat the black/white evolution there is good reason to suppose that process will lead to the same endpoint. Do the same for any neutral mutation, though, and there is no reason any given neutral mutation will have the same fate every time. For a neutral mutation, its fate has nothing to do with its effects on fitness, as there are none. Evolution, then, does not have to be owing to selection forcing the change via differences in survival and reproduction (black/white differential survival); it can be just owing to chance bobbling about.

A key question, thought Kimura, is how the fate of neutral mutations bobbling about depends on population size. In the mathematics of gas particles, this means a taller room and hence more bobbling about, more chance moves up and down. The height of the room is the analogue of population size. Mammals often have much smaller populations than, say, insects or bacteria, just because species with large bodies have fewer individuals. In a room with a really low ceiling (small population, large-bodied mammals), for example, the new particle close to the floor needs relatively few lucky breaks to get from near-floor to ceiling. In a really tall room, it needs loads of lucky breaks. It doesn't need to be lucky all the time—we expect it to bobble about, going up and down. So chance sampling is more likely to lead to a species difference when the population size is small—like us. So far, so good, and I expect that rather went with your intuition: chance events will be more influential in small populations.

Kimura could be more precise. To see this, let's stop talking about gas particles and instead think about a staircase game. The principles are the same, but the staircase game might be an easier way to capture some insights. The number of steps is the population size, and your relative position on the staircase the frequency of a mutation. If there are 100 steps and you are on step 20, you are like a mutation at frequency $20/100 = 0.2$ in the population. What are the rules of the game? First, over time you will move up and down steps randomly by a coin toss. Heads you go up one, tails you go down one. The second rule is that the game stops when you arrive either at the floor below or the floor above. If the new mutation arrives at the floor below, this is equivalent to the new mutation being lost by chance—the population goes back to its old pre-mutational state. If the mutation gets to the floor above, this will be a change to the DNA of every member of the species and will be seen as a difference between species (other species are playing parallel stair games). Fixation.

Importantly, I will assume my coin is unbiased: 50% chance of heads, 50% chance of tails. This is equivalent to saying that our mutation has no effect on fitness, and so its movement is owing to chance exclusively. For those who might note that a coin could land on its side, I say, you are right—there is a 1 in 6,000 chance of this—but let's assume heads or tails are the only options.

Because our coin is unbiased, our first insight in the behavior of this game is that if you are on step number x, then your chance of going to step $x + 1$ is the same as your chance of going to step number $x - 1$ in any one round. If you are on step 5, half the time you go to step 6, half the time to step 4. The same truth applies as you move further forward in time, such that if you find yourself halfway up the staircase your chance of eventually ending up on the floor above is exactly the same as your chance of going down to the floor below—i.e., half the time you will end up at the floor above, half the time at the floor below. Hold that thought.

Now let's run the game to mimic the evolution of a new neutral mutation. As mutations are rare when they first arrive in a population, you start on the first step. Let's see if we can figure out what your chances then are of eventually getting to the floor above.

We start by flipping our coin. As you started on the first step and half the time you throw a tail, half the time the game is over before it really started, as you have moved down to the floor below. The new mutation is immediately lost from the population—this is why the riskiest time for any mutation is when it is initially rare. But half the time you toss a head and move to the second step. Toss the coin again and you can move up to step 3 or down to step 1. Toss again, etc.

What now happens? In principle, it looks like a hard problem to work out the chance of eventually moving to the floor above—the new mutation is fixed in the population—or moving down to the floor below—the mutation is lost from the population, either immediately (the first coin toss was a tail) or eventually.

We start with the thought that if you are halfway up the stairs, your chance of getting to the floor above is the same as your chance of going to the floor below, because coin tosses are unbiased (50:50). So what then if there were only two steps on our staircase? You enter onto step

1 and half the time go to the floor above, half the time to the floor below. You were halfway up when you arrived on the first step so the chance is 50% of going from first step to the floor above. Another way to write this is to say that with 2 steps your chance of arriving at the floor above is ½. What if there are 4 steps? If you are halfway up (step 2), your chance of going to the top floor is then 50%. So the chance of getting to the floor above is ½ times the chance of getting to step 2. But we just saw that your chance of making it to step 2 is 50%. Your chance of going from step 1 to the top floor with 4 steps must then be ½ × ½ = ¼. So ¼ of the time you will start out on step 1 and make it to the floor above if there are 4 steps—solely by chance.

And what if there are 8, 16, 32, 64 steps? So the logic goes on. If the chance of getting to step 4 is ¼, then the chance of getting to the top of an 8-step staircase is the chance of going from the 4th step to the top (i.e., ½) times the chance you made it to the 4th step (i.e., ¼), so ⅛.

There is a pattern. If you start on the first step, then your chance of making it by chance to the upper floor is 1 divided by the number of steps: 2 steps, ½; 4 steps, ¼; 8 steps, ⅛; 16 steps, ¹⁄₁₆; etc. Or, more generally, the chances of going from 1 of N to all of N by chance bobbling about is $1/N$. As we happen to have two copies of every gene, if there are N of us the chance that a new mutation (just one in the whole population) makes it to fixation—all copies of the same bit of DNA become this new version—by chance alone is $1/2N$. Species like us have $2N$ steps each new mutation must have climbed to be a fixed difference between us and chimps. This is one of Kimura's central results and captures the verbal logic we saw above—that bobbling about is more likely to lead to fixation (going from rare to common) when population sizes are small (low N) than when they are large.

And notice that to get from the first stair to the top, you don't need luck always to run your way. You don't need all heads to get there. You could go steps $1 \rightarrow 2 \rightarrow 3 \rightarrow 2 \rightarrow 1 \rightarrow 2 \rightarrow 3 \rightarrow 4 \rightarrow 3$, etc. You are bobbling about randomly, going up and down steps. But if ever you step off the first step (step 1 to the floor) then you are out of the game, as you are if you are lucky enough to make it to the floor above.

This is the fundamental mode of operation of neutral evolution. Neutral evolution is predicated on the notion that the chance of going up or down the steps is equal, owing to chance alone. Selection is different. You could think of selection as adding an extra force: wind in your face pushing you down the steps—purifying selection tending to remove a new mutation—or a wind behind you tending to push you up the stairs to the next floor, which would be positive selection on a beneficial mutation, tending with each generation to increase in frequency (to head up the stairs).

So far, so good. If you thought that neutral evolution must be much more important in small populations because chance bobbling is more important in small populations, you would be right. And wrong.

And this is where the new deep truth comes about. If we are talking about mutations that have *no* effect, you would be wrong. If you are thinking about mutations that have a *small* effect, that are nearly neutral, you would be right. This is not obvious.

If we think about strictly neutral mutations, Kimura pointed out that you would be wrong. How on earth can that be true? Surely randomly bobbling must be more influential in small populations. Didn't we just see that! We just saw that the chance of going from rare (step 1) to the floor above goes down the longer the staircase, our $1/2N$ result. Well, yes, there is nothing wrong with this part of your insight. It is the missing bit that matters. Kimura, you see, noticed something else. We are not just interested in the fate of this one new mutation. We want to know the overall rate of evolution of neutral mutations, not just the fate of one.

The sampling process then tells us that when our new mutation appears in the population it has a $1/2N$ chance of getting from step 1 to the floor above just because of bobbling about. But how often do such mutations come into the population? Well, we know that too. This is dependent on the mutation rate.

Compare your DNA with that of your mum and dad and you will see changes—between 10 and 100, usually. These are new mutations, unique to you. One measure of our mutation rate is then just this average number: about 50 new mutations (per newborn, i.e., per generation). A more commonly used measure is this same number but scaled to the amount of DNA, just because with more DNA you expect more mutations. We usually call this number μ (pronounced "mu"). This then is the average number of new mutations in each newborn (i.e., each generation) for any given base pair of DNA. For any given gene, the comparable rate would be just $\mu \times$ number of base pairs in the gene. We really just scaled the mutation rate from a per-base pair rate to a per-gene rate, so call this μ per gene, rather than μ per base pair.

Kimura's key insight was that, at any given gene, the rate at which new mutations keep appearing in the population is this rate (μ per gene) times the number of copies of this bit of DNA in the species concerned. As we have 2 copies of our DNA—one from mum, one from dad—if there are N individuals in the population, there are $2N$ copies of each gene, all of which can potentially mutate. So in our population, new mutations arrive in our focal gene at a rate $2N$ times μ per gene.

What Kimura had noticed was—in retrospect—a bit obvious: the bigger the population (the higher the value of N), the more opportunities there are for new mutations to appear to undergo the bobbling-about process at different locations in your DNA: more mutations playing parallel staircase games.

And now we find a result that is odd: the net rate of evolution of perfectly neutral mutations across any given gene owing to bobbling about has nothing to do with the population size (N).

You can see this mathematically. The net rate is the rate of introduction of new mutations ($2N$ times μ) multiplied by the chance of going from rare to common by bobbling about $(1/2N)$. Mathematically, the two $2N$ terms cancel out. So the rate of evolution of strictly neutral mutations is simply the mutation rate. Wow. This was not initially obvious.

Put differently, the higher chance of fixation because of bobbling about in a small population is *exactly* canceled out by the lower rate at which new mutations come into the population. The only thing that determines the rate of neutral evolution is the rate at which new mutations appear. If you thought chance processes must be more important in small populations, this is why you are wrong. The staircase sampling process is more important when populations are small; the rate at which you play the staircase game is also lower.

The Great Evolutionary Roulette Wheel, or, Why Population Size Matters After All

Why then might you also have been right? Here we meet possibly one of the most profound evolutionary insights of the last fifty years or so. What we see above is true for *strictly* neutral mutations—these make absolutely no difference in fitness and have an absolutely equal chance of bobbling up as bobbling down in frequency. But what if we relax our assumptions a bit? What if our new mutation causes a tiny reduction in fitness? Does the same logic still apply? Surely a minuscule effect on fitness will not matter? A mutation in a pseudogene might have a tiny, tiny effect. For example, G and C nucleotides are ever so slightly more expensive to make (they have one more nitrogen atom) than A and T. Could a mutation from A to G that slightly increases the costs really be subject to natural selection because of its teeny-weeny effects on your chances of surviving? This is quite a profound question: What are the limits of selection and in turn the limits to perfection?

This extension of Kimura's work was done by his colleague, Tomoko Ohta, who in 1973 proposed the so-called *nearly neutral theory* of evolution. While the neutral theory considers new mutations that have no effects on fitness, the nearly neutral theory extends this to consider the possibility that mutations might have minuscule effects. Her work provides the conceptual foundations for our understanding of much of the oddness of our DNA.

Both the neutral and nearly neutral theories assume that bobbling about randomly is a key part of evolution. Nonetheless, while it might be called the nearly neutral theory, in terms of its predictions it could not be more different from the strictly neutral theory that we saw above. The nearly neutral theory, you see, predicts that selection will be less influential when populations are small. The rate of evolution will then be higher owing to bobbling about when populations are small, as the common slightly harmful mutations cannot be eliminated as often. The neutral theory, by contrast, says—as we just saw—that the rate of evolution is independent of population size and just depends on the mutation rate. As I said, the two models could hardly be more similar in their assumptions—they both think bobbling about is the key process—and more different in their predictions.

The idea that we, like all species with small effective population sizes, cannot purge our DNA of slightly harmful mutations is what underpins the new view of our genomic imperfection. This is why we end up riddled with bullet holes in the places that don't majorly matter, wing tips and fuselage.

Why, you might ask, would such a small modification to the theory make such a difference? We can think again about our staircase game. Again, we start on the first step and ask how often we will get to the floor above. What is now different is that there is a weak wind in our face pushing us down the steps. An alternative—but equivalent—way to think about this is that our decision is based on the spin of a roulette wheel. These two ways of thinking about it are the same, as they presume that it is just a little more likely that you will go down a step than up a step. Let's run with the roulette wheel model.

A European roulette wheel has 18 black numbers and 18 red ones. And 1 green one (number 0). Put your money on black to win. On the average, in a real game of roulette you lose, as, while there are as many reds as there are blacks, there is also this one green. So, if ever you pick either red or black, the probability of success is just a little under 50:50 (18/37, to be exact—i.e., 48.6% of the time you win). You will on average lose. There is a reason the green is there. The neutral equivalent would be the same roulette wheel but without the green. If we changed the (still equal) number of red and black holes but kept one green, this would be equivalent to changing the extent to which a mutation is a little bit deleterious: the more reds and blacks, the closer your chances will be to 50:50, and the weaker the wind of purifying selection in your face.

Now, rather than "heads you go up a stair, tails you go down one," we will have "black you go up, red or green you go down." This mimics our slightly deleterious mutation: it doesn't quite have a 50:50 chance of going up one step each go. What we are interested in is the dynamics of how the outcome of the staircase game varies depending on the number of steps, our equivalent of the population size.

The rate of mutational input will be the same, $2N$ times μ: both models assume the same chance of entering a staircase game. Where the neutral and nearly neutral models differ is in the half of the calculation concerned with whether a new mutation makes it up all the steps. Remember from before a key result of the neutral model: that the probability that our new mutation gets to fixation is $1/2N$. This we got to by asking about the chance, if you are halfway up the stairs, that you will get to the floor above as opposed to the floor below. We saw that both futures are equally likely if we are simply tossing an unbiased coin. In turn, from that result we could see how, if you started at step 1 (a new mutation in the population), your chance of going up to step 2 was 50% and your chance of being immediately lost (going to the floor below) was 50%. In the roulette stair game this isn't true. Your chances of going from step 1 to step 2 are 48.6%. So if there are just 2 steps your chance of getting to the floor above is 48.6%, not 50%. This captures two important truths: mutations that are a little bit harmful can still get to fixation (the top of the stairs) but have less of a chance of doing so than a neutral mutation.

The realization that bobbling about can enable fixation of a slightly harmful mutation is an important result but is not Ohta's key result. What matters is how your chance of getting from the first step to the top of the stairs depends on the number of stairs. What she finds is that the more stairs there are (the larger the population) not only is the chance of getting to the top of the stairs reduced, as it was with the neutral theory/coin toss, but it disproportionately declines as the number of stairs goes up. This means that in large populations the gentle wind in your face is enough to stop you getting to the top of the stairs, but the same gentle wind has less effect when the population is small. Small populations are disproportionately more prone to fixing slightly deleterious mutations. In large populations even weak selection is good enough to keep the DNA mutation-free.

To see this, now suppose you made it to step 2. What is your chance of getting to step 4, the top of a 4-step staircase? Remember, in the neutral model the chance of going from step 1 to step 2 is the same as the probability of going from step 2 to step 4 in the 4-step game, is the same as that of going from step 4 to 8 in the 8-step game, etc. More generally, if at step x you are halfway up, your chance of going from step x to step $2x$ (the top floor) is the same as the chance of going from step x to step zero (the lower floor), and doesn't depend on the value of x. Extend that logic and we find that your chance of going from step 1 up all $2N$ steps was $1/2N$. This is not true when we are playing the roulette wheel of evolution rather than the coin toss.

Playing the roulette wheel version of the stair game, the chance of going from step 2 to step 4 must also be less than 50:50. Importantly, however, it is less than the chance of getting from step 1 to step 2. Similarly, your chances of getting from step 4 to step 8 are lower than your chances of getting from 2 to 4. This perhaps not-so-obvious result is key to understanding Ohta's breakthrough. It indicates that deleterious mutations can still reach fixation by bobbling about, but that with a smaller population (fewer steps) this probability is not simply higher, but disproportionately higher.

Perhaps this is an easier way to intuit this result: let's suppose you come to a roulette table with a certain amount of money in your pocket: perhaps $10 or $100 or $1,000. You bet $1 on black each time and receive $1 if you win (along with your original $1), nothing if you lose (and you lose the $1 stake). You will end the game if you run out of money or if you double the money you started with. This is like starting halfway up the staircase.

If you start with $10, do you have a higher chance of doubling your money (versus going home empty-handed) than if you started with $1,000 playing $1 each bet? If you are playing the coin toss game, the answer is no: half the time you leave empty-handed, half the time you double your money no matter what amount you start with. If you are playing roulette the answer is yes: if you start with only a little money, you are more likely to double your money than if you start with lots of money (no matter what, you are still more likely to go away empty-handed than to double your money). The chance of doubling your money if you start with $5 is less than 50%, but not too bad—a couple of lucky breaks. The chance of going from $50,0000 to $100,000 playing $1 bets each time is negligible.

I did a simulation of playing a roulette wheel to get some better numbers. Assuming your chances of winning on any turn of the wheel are 48.6%, if you start with $5, a bit over 40% of the time you can double your money if you stop as soon as you double your money (reach $10). Start with $50 and stop as soon as you either get to $100 or lose all your money, and you win on a roulette wheel about 5% of the time with a $1 bet each time. Getting from $100 to $200 is rare, around 0.5% of the time. If your chances were 50:50 in any coin toss, these probabilities would all have been the same, all 50:50. That green slot on the wheel really matters. Casinos know this about roulette wheels, so when you are on a winning streak, they entice you with free drinks to keep playing—the more you play the more likely it is you will in the end lose (end up on the ground floor). Don't spend your life on a roulette wheel. If after a few spins you are lucky enough to be up, then quit.

What does this do to our overall calculation of the rate of evolution? As we noted, we assume that the rate of input (new mutations) is the same in the neutral and nearly neutral models. However, mutations that are a little bit bad for you are disproportionately more likely to end up on the floor below (sequence conservation, no evolution) when the population size is larger. Unlike in the neutral model, the increased net rate of new mutations arriving in large populations (remember the $2N \times \mu$) doesn't fully offset this lower chance of fixation in large populations. The net rate of evolution will then be lower in large populations, as they are more efficient at not fixing mutations with the odds slightly against them. This all comes about because, if the odds are against you going up a step, you are less likely to get to the top of a long staircase than a short one, even if you are already halfway up.

How does this work out in practice? Are mutations that are ever so slightly deleterious actually affected by this? In the nearly neutral framework we often think of mutations as being one of three flavors. On the one hand are mutations of such a small effect that their fate is indistinguishable from that of a strictly neutral mutation. This sort of makes sense. For any population size, some changes must be so tiny that selection cannot discriminate between the two versions. In the stair game, these are equivalent to playing a roulette wheel with millions or billions of reds and blacks and only one green. Your chance of success is really close to 50:50. These have, Ohta predicts, an evolution rate extremely close to μ, the mutation rate, the same as for strictly neutral mutations. Consequently, we call these "effectively neutral" mutations.

A second class are those mutations that are so deleterious that they effectively stand no chance of getting from rare to common by bobbling about. These are playing a game where the chance of being able to increase in frequency each generation is too far below 50:50. These are just simply deleterious or harmful mutations. Bullets that hit a plane in the engine would be included in this group, but so might more subtle mutations that bobble about a bit but are eventually removed from populations.

Then there is a middle ground. These can go from rare to common by bobbling about, have a chance each generation close to 50:50, but not so close as to be effectively neutral. They can go to fixation by bobbling about, but their net evolutionary rate is a bit below the mutation rate. These are referred to as weakly deleterious or slightly deleterious (both terms are used interchangeably).

Just how close to 50:50 do you have to be to be effectively neutral as opposed to slightly deleterious or fully deleterious? The key result of Ohta's calculations is that for any given mutation, which of the three boxes it sits in (i.e., the ratio of reds/blacks to the one green) is dependent on the population size, for reasons that we just saw. Tiny-effect harmful mutations we expect to be eventually removed from large populations (more steps) more often than from small ones. Large populations have a better chance of being perfect and holding on to perfection by removing even small-effect mutations. Small populations, by contrast, have a problem getting rid of small-effect bad mutations. We end up with lots of bullet holes in relatively unimportant places.

The diffusion theory math allows us to put some numbers to this. There are two numbers that really matter. First is the number of individuals within a given species. To be more exact, it turns out that the number that really matters is not the head count of individuals but, instead, something called the *effective population size*. If only some males get to mate, for example, the head count and the effective population size are not quite the same thing. To a first approximation this is a technical concern, but for the sake of accuracy, we will refer to this number, called N_e. There is a lot of chatter in evolutionary circles about N_e.

The other number that matters is the extent to which our new mutation reduces fitness—the strength of the small breeze in our face or the number of red/black slots compared to the one green. This we call s, a measure of the strength of selection against the mutation compared with the other version of the same bit of DNA.

For a single-celled species, you can imagine measuring s in the lab. You take one strain of your organism without the mutation and let it grow from small numbers to large numbers (this is not staircase climbing, by the way, this is just population growth—changing the number of stairs, if you like). You then put a number on the growth rate: how many cell divisions per hour, perhaps. You do the same for exactly the same organism except that it has the mutation. Comparing the two rates—with and without our mutation—we can see how the mutation affects how fast the cells can divide. In experiments in the lab, we can now measure small differences of the order of about a 0.1% difference in growth rates. I think that is amazing, but we cannot for the most part directly measure the sorts of differences in fitness that the nearly neutral model considers at its lower end. A small s could be very close to 50:50, perhaps 49.9999999 to 50.0000001, far less than a 0.1% difference.

What Ohta's theory then does is say that for any harmful mutation you might want to consider, which of the three bins it sits in depends on the effective population size. A mutation will be effectively neutral if the cost of bearing the mutation (s) is much less than $1/(2N_e)$. (The 2 in this statement again comes about because we have two copies of each of our genes.) Conversely, a mutation that causes a fitness reduction, s, much greater than this ($s>1/2N_e$) is a deleterious mutation with no chance of going from rare to common and getting fixed. You also stand almost no chance of doubling your money on a roulette wheel if you start with lots of money and bet $1 a time. In between are the slightly deleterious mutations that reduce fitness by about $1/2N_e$. These have a chance of getting to be fixed but a bit lower than μ, the rate for effectively and strictly neutral mutations.

That is all a bit abstract, so let's put some real numbers onto these estimates. We think for humans our effective population size is about 10,000–20,000. Many people when they first encounter this number think that it is just plain stupid! There are billions of us (and growing). Well, yes, there are, but when we evolved the differences that we see between us and chimps, our closest relatives, the human population size was much smaller. The current billions of us are too evolutionarily recent to make any meaningful difference. What this then means is that a mutation that reduced fitness by about one in a million will sit in the "effectively neutral" class for us. This is because for humans $1/2N_e$ is about $1/30,000$, i.e., about 0.0003. If a mutation's effect on fitness is much less than this, then it will be effectively neutral. To be slowed or prevented from getting to fixation, it needs to affect our fitness by reducing it by 0.0003 or more if on average we have a fitness of 1. Conversely—and here is the key point—a mutation with the same effect on fitness in a bacterium with a population size in the billions would sit in the class "deleterious" without a hope of getting to fixation. Many species of bacteria can clear out deleterious mutations from their populations much more effectively than we can, just because their effective population sizes are massive. Mutations that reduce fitness from 1 to 1–0.000000001 will still be "visible" to selection and be expelled from the population. I still find it hard to grasp that selection in bacteria can be that efficient. But it does mean that selection is amazingly efficient in species with large populations and rougher-grained in others such as us.

The net effect of all this is that more mutations are likely to be in the effectively neutral or slightly deleterious class when populations are small. Mutations in these classes can bobble about and get to fixation. For this reason, species with small effective population sizes—like us—can evolve faster than much more populous ones. But this faster evolution is not progressive evolution—it is decay. Put differently, a bullet at the wing tip having a small effect will eventually bring the bacterial plane down, but not our plane. We accumulate these small holes.

But what has this got to do with our genome? We often think of mutations as a change in DNA replacing one base pair with another. A C-to-G change, perhaps. But the nearly neutral theory makes no presumptions about what a mutation is. It applies to any change to our DNA that starts out rare in the population. Think of mutations not just as changes between one nucleotide and another. Think also of gain of bits of DNA that do nothing for us. Like other big-bodied species with small populations, we were unable to do much to stop these going from rare to common. Does this explain the apparently imperfect anatomy and behavior of our genome? As we shall see, it explains why our DNA is full of dead jumping genes (virus-like bits of DNA that autonomously jump about the genome, also called *transposable elements*), while the DNA of bacteria is nearly all protein-coding. It can also explain why so much of our DNA is active but not obviously functional. This is the subject of the next chapter.

5

The Nearly Neutral Theory and the Problem of the Bloated Genome

Tomoko Ohta has given us a profound truth. When population sizes are small, selection against mutations that harm us a little is less effective. While physically minuscule species with massive population sizes can purge their genomes of new mutations that are a little harmful, we cannot. The evolution of imperfection is—for us—an inevitability. But what has this to do with the size of our DNA or the way our cells function?

We need to expand the notion of a mutation from a single base-pair change to other sorts of changes, like gain or loss of DNA. Any gain or loss event will affect one individual bit of DNA in the first instance. It will be a rare mutation, just a rather more complex one than a change of one base pair. As we shall see, gains that are just a little bit harmful can explain why our DNA is bloated while yeast and bacteria have the lithest genomes. Our DNA is full up with rubbish and functions badly. If our cells are cities, they are chaotic ones with cars running the lights, broken hydrants, and traffic accidents aplenty.

Not All Rubbish Is Trash, Some Is Junk

At this juncture we need to classify our rubbish. On the one hand there is rubbish we put out for refuse collectors. They come around in big trucks, take away our stuff, and put it into landfills or incinerate it. In the US this, I believe, is called garbage or trash. There is another sort of rubbish, the stuff we don't use but just accumulate in lofts and garages. This is junk. This we don't send to landfill, as it isn't doing much harm. It isn't the decaying remains of last week's dinners. The idea we are concerned with is that our DNA is full of junk. Genomic trash is different. These are the mutations that get expunged from populations because they are bad, removed (eventually) by the great trash collector that is purifying selection.

How does this relate to genome bloating, the tendency for our genome to have big introns and lots of DNA between genes that we saw in chapter 2? Consider the fate of a duplication of a bit of unimportant DNA, or an insertion of a bit of DNA that doesn't really disrupt anything. Jumping genes are doing this all the time. This stray insertion may well reduce fitness a small degree—we need to spend energy copying it every time a cell divides; it may disrupt the way nearby genes are expressed. But there is a good chance that it will just be associated with small effects. And there is the rub: it may well be too selectively unimportant for a large-bodied species with small population sizes to do anything about. In effect, we just throw it into our DNA garage or loft and forget about it.

By contrast, bacteria, yeast, etc., are single-celled, have massive populations, and so have highly effective selection to perfect their genomes. In their lithe, uncluttered lives, bacteria and yeast send what we think of as junk to landfill. Thus, according to the nearly neutral model, a rubbish- (i.e., junk-) bloated genome is just what we should expect in a species like us that has a big body and ancestrally small population size.

This single idea has the potential to be a powerful unifying theory. If we look across all of eukaryotic life (the group that, unlike bacteria, have a nucleus in their cells), from single-celled species like yeasts to big creatures like us, one regularity falls out. As Mike Lynch, now at Arizona

State University, noticed, the effective population size predicts such things as number of introns, size of introns, size of intergenic space, and proportion of the genome that is dead or decaying transposable elements (those virus-like genes that jump about). These all go in the same direction: as effective population size goes down, all independent measures of the amount of what looks like genomic junk go up (see, e.g., fig. 5.1). We have a large genome, large introns, lots of introns, a big distance between our genes, lots of dead and decaying jumping genes. Single-celled yeast does not. We are hoarders of DNA junk, they are not.

The trends are not absolute, and in some sub-lineages (e.g., some insects) we find intron density (number of introns per base pair of coding exon) is unchanged while mean intron size is fluid. There are also some oddities such as whales, with very low population sizes, having introns that are a little smaller than ours, when we might have expected the opposite. There are also energetic concerns that intervene. In flying vertebrates (bats and birds), introns tend to be small. This probably reflects the energetic demands of flight, making selection on intron length in these species stronger, as they need a tighter rein on energy usage. Overall, however, we see that many of these measures both co-vary with each other and correlate with effective population size. While the data is noisy, larger-scale analyses, allowing for the fact that related species are likely to be alike, still report the same trends: introns tend to be larger when populations are smaller. It all fits with a model in which the key problem our DNA has is getting rid of the sort of rubbish that only weakly affects us. We end up filling up the loft and garage with unwanted but untrashable stuff.

The model can also explain some of the variation in intron sizes that we see between genes in our genome. The most highly expressed genes in our genome have smaller introns, probably because selection on them is a bit stronger to stay small. Transcription of long introns in highly expressed genes would be especially costly. Insertion of superfluous DNA into these genes is thus more likely to be in the simply delete-rious class than the same insertion in rarely expressed genes. This isn't all there is to intron size, however. There can also sometimes be se-lection to keep an intron long. This could, it has been suggested, enable

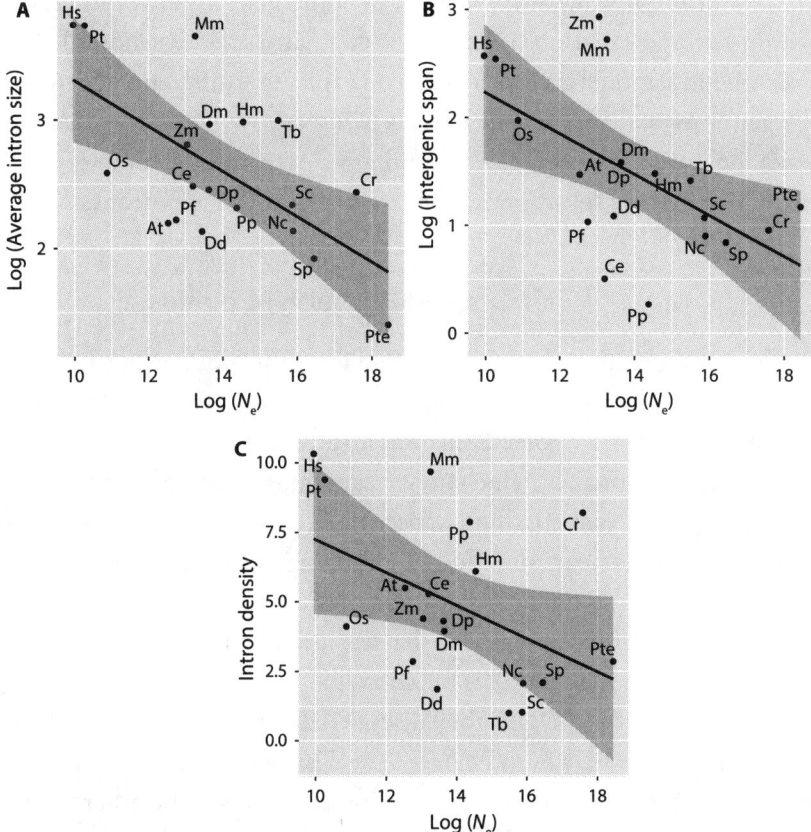

FIGURE 5.1. **Intergenic span, intron size, and intron density are higher when effective population size (N_e) is low.** Average intron size is the mean size of introns; intergene span is the total span of DNA between coding genes; intron density is the mean number of introns in genes that have introns. Labels of data points refer to first letters of the genus and species name: Pp = *Pristionchus pacificus*, Ce = *Caenorhabditis elegans*, Dp = *Daphnia pulex*, Hm = *Heliconius melpomene*, Dm = *Drosophila melanogaster*, Mm = *Mus musculus*, Pt = *Pan troglodytes*, Hs = *Homo sapiens*, Cr = *Chlamydomonas reinhardtii*, Os = *Oryza sativa*, Zm = *Zea mays*, At = *Arabidopsis thaliana*, Tb = *Trypanosoma brucei*, Dd = *Dictyostelium discoideum*, Pte = *Paramecium tetraurelia*, Pf = *Plasmodium falciparum*, Sp = *Schizosaccharomyces pombe*, Sc = *Saccharomyces cerevisiae*, Np = *Neurospora crassa*.

control of the timing of expression of some genes. However, this is within genome variation (i.e., between genes in a species genome). The nearly neutral model is more focused on explaining differences between the genomes of different species. This being said, we also think that the efficiency of selection varies around a genome, and in some species the regions where selection is more efficient have small introns.

Most of Our DNA Is (Probably) Pointless

Perhaps the greatest challenge to the view that most of our DNA is pointless junk put into our genomic garage, however, came from the so-called ENCODE project. This was an attempt to know what in our genome is functional, by which they meant doing something in the cell. They considered that anything transcribed (ever), that binds to something else, that is near something interacting with something else, is "functional." By this definition a striking 80% of our DNA was deemed functional. They thus announced that the idea that most of our DNA is "junk" was dead. It nearly all is functional.

Almost 75% of our DNA, they discovered, is transcribed into RNA. The true number may even be over 90%. This was a surprise. Of these transcripts, however, only 4% are protein-coding ones. The others make non-coding RNAs. Thus, they surmised, there must also be a deep hidden world of functional non-coding RNA.

Who, then, is right? You might think that this is just a niche tussle that can be confined to academic ivory towers. This, however, isn't an ivory tower issue. If we are to understand the causes of genetic diseases and do something about them, then we need to know what is functional in the sense that if it changes it harms us. Similarly, can I insert a new gene to make someone better almost anywhere in their DNA? Or will I break important functions if I do this, making the patient sicker?

This 80% headline figure raised more than a few eyebrows in evolutionary circles. The nearly neutral model predicts that very little of our DNA is doing something selectively important. It is made up mostly (two-thirds) of old jumping genes, viruses that have incorporated, etc. These things, according to Mike Lynch and colleagues, are there not because they are useful, but just because we cannot get rid of them. How can we square this circle?

To some extent the two camps are at cross-purposes, as they simply start with different definitions of what it means for something to be "functional." ENCODE considers it defined by cellular activity, the nearly neutralists by whether it shows evolutionary hallmarks of DNA that are conserved because they do something important for us (as in the Ka/Ks test).

I have little sympathy, however, with the ENCODE worldview. My problem is this. Whenever we see some sort of molecular interaction (molecule A bangs into molecule B), we could be looking at an adaptation: A banging into B is necessary for the cell. However, if you see a pedestrian thrown forward onto the road by a car while crossing the street, you would, by this logic, then have to suppose that the adaptive function of cars is to throw people in front of them up the road. According to ENCODE, people in the vicinity of the accident also are functional for this interaction, just because they are near. Under the ENCODE definition, whether accident or not, the car–pedestrian interaction has a function. The evolutionary folks didn't like this and neither do I. Functional in evolutionary books means not just doing stuff, but being under selection to do stuff—not a strange accident.

The strange accident idea may seem odd but has a solid experimental basis. Imagine this experiment. Take a gene coding for a protein that causes a cell to glow. This allows you to see (with your eyes, under a microscope) whether the gene is working. We call these "reporter genes," as they report to us whether the process of transcription and translation is working. Now add a string of random DNA in front of the reporter gene. The DNA in front of a gene is part of its on/off switch mechanism. If you now add this combination of random DNA and reporter gene to a cell how often would the cell glow? That is to say, how often would a random bit of DNA function as an on switch? Remember that this is random DNA—it hasn't evolved to allow a protein to be expressed. If DNA accidents are common, we might commonly see the cell glowing. If DNA accidents are common, we should not presume that everything that happens, happens because it is the product of natural selection.

This sort of experiment (or variants thereof) has been done in bacteria, yeast, humans, and flies. And in all cases, we see that the random DNA very commonly functions as an on switch. In one experiment, for example, lots of random bits of DNA 120 base pairs long were introduced into yeast. About a half resulted in the production of the glowing protein. If you put human DNA, or reversed human DNA, into yeast cells, more than 99% of it is transcribed. The real switches in DNA are usually better at doing the job, but some expression is common. For the RNAs that come from the DNA between genes in yeast, in just about all cases we cannot exclude the possibility that they are just spurious transcription.

We now think that the evolution of gene expression is as much about suppressing this unwanted transcription as it is about enabling the wanted. If you compare the profile of yeast's real (evolved) gene expression with that of the naïve DNA, the evolved genes have higher expression, fewer bases are transcribed, and there is less transcription from both strands of DNA. This suggests that yeast has evolved to suppress random expression to enable clean high-expression levels of the transcripts that really are functional.

Similar experiments have been done to see how often random transcripts might be spliced. The spurious RNAs from human DNA put into yeast are, for example, spliced about as often as yeast's own genes. In all cases, activity is observed that ENCODE would define as evidence for functionality. I would suggest that accidents happen.

While these sorts of experiments tell us that lots of biochemical activities could be random accidents, what we really want to understand is what in our evolved genome is functional in the sense that if a mutation were to change it, the organism could be better or worse off. The mutation could be a change to a single nucleotide or loss or gain of blocks of DNA. The latter could include the insertion of DNA, for example of a virus, into my DNA.

This is a question we can approach from several angles. One way is to do a genetic experiment and take the possibly functional thing away. DNA removal like this is called a knockout. When it comes to genes, we can also leave the gene intact in the DNA but stop it from being transcribed into RNA. We can make knockouts for individual genes, for example, to see what function they might have. You examine the organism with the gene and without the gene (or its transcriptional product) and do a spot-the-difference competition, or try to estimate s, the strength of selection on the deletion, by laboratory growth experiments, as we saw in the previous chapter.

The problem with this approach is that while it can offer evidence that something is functional, a negative result is quite likely to be misleading. We can do this in yeast, for example, where all the signs are that its protein-coding genes are needed. They almost all have Ka/Ks <1, indicating that selection is acting on them. But in the lab many gene knockouts have no obvious effects, or just modest effects. Can we conclude that the genes in question aren't functional despite the evolutionary evidence?

The answer is a resounding no. One problem is that we can't measure evolutionarily relevant fitness effects. The differences could be too subtle to be spotted, but still important. Remember selection can operate on mutations of very small effects in species with a large population. In the lab we can't measure subtle selection.

We also know that we have a more general problem with so-called false negatives: genes that appear to have no effect on being removed, but do have a function. A few years ago, I looked at this problem in yeast. We asked why so many genes could be removed, one at a time, from yeast, and the resulting cell do just fine. We found a simple answer. Most genes that, when knocked out in the lab, appear to have no function are simply not needed in the laboratory because of the indulgent conditions in which we grow the yeast.

The "non-essential" genes often code for the ability to eat a certain food that isn't in their lab diet. Or, the endpoint of the metabolism is being supplied in their food. This is a bit like taking an organism that can synthesize vitamin C, giving it loads of vitamin C, and then wondering why, when on this diet, a cell with a *GLO* knockout is just fine. You don't need to make vitamin C if you are given loads of it. Were I to send you somewhere sunny and dry and you then decided, as you are fine with or without your raincoat, that raincoats have no function, you would make the same mistake.

Not having an obvious function could then just mean "no utility in the current conditions." Other experiments in yeast show that you can find other conditions where a gene's absence does have an effect. But the problem remains: just because there is no effect when you take something away doesn't mean that the thing taken away is always useless. This approach also has the problem that the very act of knocking a gene out could be what causes any effect seen—these are false positives. For some approaches we know this is often an issue.

Potentially, then, a better way of working out whether a bit of sequence really has a selectively important function is to look to see how much is preserved over evolutionary time, as preservation is unlikely if the DNA has no function and the activity is accidental. This is why Ka/Ks <1 for most genes—the amino acids are preserved over evolutionary time more than expected given the background rate of evolution. Different folks have tried to estimate this a few different (independent) ways for our genome as a whole, not just the protein-coding genes. You can look at the extent of sequence conservation, or you can ask how much selection could realistically be operating. We can also look for evidence from the frequencies of mutations in populations. The premise here is that if selection is operating, mutations will be mostly low-frequency, mostly toward the bottom of the steps (they have selection's wind in their faces). No matter what the approach, it always comes out with a very similar estimate: somewhere between a few percent and about 15% of our DNA is under selection to preserve its utility. Around 10% of our DNA being functional is a good estimate. It never gets close to 80%.

The methods aren't perfect, and many have a classification issue with mutations in places in the genome where a certain amount of DNA needs to be there, but it doesn't matter what the identity of the DNA is. For example, we have seen how GGC, GGT, GGA and GGG all code for the amino acid glycine. Here would be a case where we absolutely need a nucleotide at the third position (A, C, T, or G), but it may not matter which of the four is employed: presence, but not identity. While this sort of selection for presence rather than identity certainly holds, we would see selection against mutations that remove the third nucleotide even if we don't see selection for the *identity* of the third nucleotide. It is far from clear that such effects can explain why, for example, our introns and intergenic DNA are longer than those of mice.

The Behavior of Genes: Small Populations and Error-Prone Gene Expression

The nearly neutral model does not just address the problems of the evolution of genomic anatomy, it also addresses the problem of genomic behavior. It can also be extended to consider how much of the activity of our genome is accidental or not. As we saw above, accidental RNA generation from random DNA seems very common. If the 10% figure for functionality is right, many other interactions and cellular activities are also likely to be cellular traffic accidents. A species like us with a bloated genome owing to small population sizes is likely to have less ability to control how gene expression works and so should have more such accidents. Are then our cells like somewhat chaotic cities in which the traffic rules aren't always obeyed? Or, perhaps, is there reason to the complexity of the biology of our cells?

To address these issues, there has been interest in what you might call alternative processing of genes. In the textbook version of gene expression, for each gene we start transcribing DNA at one place in the DNA; the RNA transcription is also terminated at one place, and a tail of A bases is added (the poly-A tail). This RNA is then often spliced in one way to make the mature messenger RNA. When the ribosome processes this, it finds the same starting ATG each time and stops the translation at the first stop codon. The protein is then correctly folded and sent off to where it is needed in the cell.

If only that was true. We often start transcription of a gene at different places (alternative transcription initiation); we start translating at different places in the mRNA; and we put loads of A nucleotides at the ends of RNAs starting at different places. When we splice our genes, we produce a whole host of different splice forms, and when we get to the end of translation we frequently run through the stop codon and keep translating. We also mistranslate (put in the wrong amino acid), mistranscribe (put in the wrong nucleotide in an RNA), misfold the protein, and send it to the wrong place. The ribosome also can slip and run on in a new reading frame. Remember that the ribosome translates mRNA in blocks of three, so ATG GGC CTG TTA gives a particular run of amino acids. Ribosomal slippage (alias *frameshifting*) means that the same mRNA is now read differently. Imagine that the ribosome jumps ahead one base after the first A. Then it would translate the message TGG GCC TGT TA. . . . This will be utterly different from the canonical protein. We also miscopy DNA through DNA replication— these being a major source of mutations.

Are these alternative events adaptations or cellular traffic accidents? Whenever I go to conferences with molecular biologists, I am always amazed at how often they assume that the explanation for all these things must be some hidden adaptation. For example, whenever the ribosome reads through the stop codon, there will be speculation about how this must diversify the set of proteins that we have. In some cases, readthrough like this is almost certainly functional. Some viruses also program this frameshifting to generate a different protein. We also expect that some alternative splicing events are functional.

But is it likely that all these events are programmed adaptations, or might they mostly be accidents? For protein folding, we are happy that we know the answer: misfolding is usually an error. We know this because misfolded proteins go to a cellular recycling machinery where they get broken up into their constituent amino acids, ready for use again.

We can also use the way sequences evolve to inform the debate. Consider, for example, the problem of translational readthrough. The mRNA starts to be translated at the ATG, which gives methionine, then runs codon by codon until it meets a stop codon. It should stop there, disengage, and release the protein. Readthrough is what happens when this disengagement does not happen, and instead the ribosome puts something else in where there should have been a stop (some amino acid or other) and then keeps translating. It may stop again downstream, or may just crash into the poly-A tail, which will be a sort of sand trap for out-of-control ribosomes. Perhaps you have seen them: on steep hills there are sandy continuations of the road to help stop out-of-control trucks.

This problem is interesting because the three stop codons are not equally prone to readthrough. Of the three stop codons, TAA is the least prone to readthrough, TGA is the worst, and TAG is somewhere in between. If the nearly neutral model is right and it is also the case that selection doesn't favor a ribosome reading through the stop codon (i.e., readthrough is an error, an accident), then we expect TAA usage should be higher in species with large effective population sizes. With my colleague Alex Ho we did this test and indeed found what was predicted. Organisms with small population sizes use less of the least-error-prone stop codon. Organisms also like to use a nucleotide just after the stop codon that reduces the readthrough rate. This indicates that when selection operates it tries to reduce the readthrough rate.

This agrees well with recent analysis of patterns of splicing. Our genes produce a veritable plethora of splice forms. Recent analysis finds that the smaller the effective population size, the richer the diversity of splice forms, as expected if selection tries to limit the diversity of splice forms. Most of these have the hallmarks of being so much rubbish: they are uncommon transcripts, they tend to change the frame of reading the genes (giving a weird protein if translated), and their splice sites (the sites where the joins are made) tend not to be evolutionarily conserved.

This all suggests that molecular events such as readthrough and alternative splicing are problems, because selection is too inefficient to stop them when populations are small. They are mostly traffic accidents, not adaptations.

This same message is reinforced when we look at variation between genes within a species. This between-gene methodology has been spearheaded by Jianzhi Zhang of the University of Michigan. He reasons that within a species, selection will be stronger on genes that are expressed more. Put differently, a driver who rarely goes out on the road is less commonly under selection not to cause traffic accidents than someone driving around all day long. The latter has so much more opportunity to cause damage.

There is lots of evidence that this assumption is right. For example, as we discovered many years ago, there is one consistently good predictor of how fast a protein evolves compared to others in the same species: how much it is used. Highly expressed, highly used proteins are slow-evolving. This is true in every species. This is just as the nearly neutral model would predict. Mutations in highly expressed genes are more likely to do real harm than mutations in lowly expressed genes. They are the regular drivers, and mutations to their cars that cause more traffic accidents are more likely to be counterselected. There is more potential to mess things up.

Incidentally, the same is true for the evolution of words in our languages. The most-used words change the most slowly over time. In English, for example, over time verbs tend to move toward the regular past tense: just put "ed" at the end of the word. We add and kill words: they are add*ed* and kill*ed*. The words slow to change, and to stay as irregular verbs, tend to be the most highly used: go and went (not "goed"), have and had (not "haved"), do and did (not "doed"). But listen to any passing three-year-olds—they will say "goed."

What Zhang finds is that all these features of alternative gene processing are less common in our more highly expressed genes. They tend to use one transcription initiation site, have one translation start site, add the poly-A tail at the same place, and not have a diversity of different splice forms. They also tend to use the best stop codon (TAA) more, and so have lower rates of translational readthrough.

The above synthesis suggests that while certainly adaptation happens, when we look at our DNA we are looking not at the perfect product of selection, but instead at the product of weakened selection. With large bodies come small populations and an inability to prevent mutations that are just a little bit bad for us from going from rare to common. Think of our gene processing as a set of busy city streets with unfortunately all-too-common errors—cars going through red lights, collisions that shouldn't be happening, people arriving at the wrong destination, cars that won't start. You name it: if it can go wrong, it probably will. Our cellular city functions—but not especially well.

The Complexity Problem

There is one somewhat glaring problem with the accident-prone view, however. This is that species with big bodies and small populations tend also to be rather complex beings. It is fine for yeast and bacteria to have streamlined genomes with one gene making—cleanly—one protein; they are single-celled beings. We not only have lots of cells, we have lots of different types of cells. Somewhere in the complexity of our DNA must also be the answer to how we develop from a single fertilized egg, the zygote, to something more complex. Our DNA must have instructions to tell certain cells to be liver or neuron, heart or kidney.

We remain uncertain about what is at the heart of this complexity. Some cell types are controlled by the generation of new splice forms of some key genes. We see this when would-be nerve cells become nerve cells: a few key changes to splice forms matter. In flies, the major differences between males and females, for example, another form of complexity, all stem from splicing a key gene one of two different ways. There is also evidence that some introns, when spliced out, go to regulate other genes. They aren't necessarily just so much junk.

Our genome is littered with the remains of old jumping genes (bits of DNA that autonomously jump about, otherwise called transposable elements) or old viruses, a bit like HIV, that have integrated into our DNA. About 8% of our DNA is indeed such endogenous retrovirus, while only 1.2% of our DNA codes for our own proteins. Generally, the larger the genome, the more these remnants of old jumping genes predominate. In a few cases we know that these old remnants of past infections have been domesticated, repurposed to do something useful. I research a few of these. We discovered that of the several thousand copies of an old retrovirus—endogenous retrovirus H—some seem to be important for our early embryonic development. Most seem to be just untranscribed, decaying fragments, however. Another old jumping gene, called *PGBD1*, appears now to help regulate how cells change from would-be neuronal cells into actual neurons.

It would be remarkable if, in the great morass of DNA that we have, we couldn't find some sections that have been repurposed. In all that junk in our garage there must be something we can use. But can we really think that when most of our DNA is the remnants of just a few jumping elements, that this is all there to enable complexity? There is a much simpler explanation: we could not get rid of most of it because selection is really a weak force in humans. Likewise, DNA seems to have an ability just to be transcribed, and many proteins stick to just about anything, so we cannot use molecular interactions to infer functionality.

Some of this debate has centered on the role, or lack thereof, of so-called long non-coding RNAs, or lncRNAs for short. These are RNAs longer than 200 bp that don't get translated into protein (although some do accidentally get caught up in ribosomes—accidents happen). We aren't sure how many of these we have; estimates range from around 20,000 to around 100,000 such genes in the human genome. As they can each be processed in a myriad of ways, there are likely to be many more resulting transcripts (numbers vary from about 270,000 to 325,000 transcripts). These are often suggested to be the hidden explainers of our complexity.

If our protein-coding exons account for about 1.2% of our DNA, these things contribute possibly around 2.3% of non-coding exons, but perhaps more. Most are expressed at low levels. In these terms, their molecular output is much lower than that of protein-coding genes. For the most part they look rather like random sequence and don't usually contain conserved motif-like structures as proteins do. They can be hard to analyze, as they are often found in only one species. Standard alignment-based methods to look at how fast they evolve can only look at the ones that do evolutionarily hang around. These tend, for the most part, to show no or little evidence of sequence conservation. You cannot do a Ka/Ks analysis for these, as they don't code for proteins (the Ka bit). You can compare their rates in the bits of the sequence left after splicing (the exons) and the sections removed, the introns. Especially for the lowly expressed ones, these two rates seem to be the same, as expected if they are irrelevant accidents. For the more highly expressed ones, there are signals that selection preserves the process of removing the introns by preserving sequences that bind to proteins that aid splicing. Blocking transcription of lncRNAs' genes rarely causes any effects on cellular growth (only about 3% do), but we saw the problem with that sort of evidence. There is little evidence that many of them contribute to human genetic diseases (one study identified 83 of 14,100 that seemed to have some sort of association).

With such evidence, many are skeptical of the strong claims that they are the hidden dark matter of the genome, the explanation for complexity. They mostly look like rather nonsensical transcripts with the occasional ruby in the rubbish.

I doubt if this is the last of this debate. Plenty of molecular biologists are trying to unpick exactly what certain non-coding DNA is doing. The oddity is that when they do close analyses like this they often find functions that look to be real functions (by everyone's definition). However, they also tend to focus on cases that look interesting in the first place.

There is some important evidence directly in favor of the nearly neutral model over the complexity model. The two ideas probably make different predictions about how fast our protein-coding genes evolve. Mutations hit these all the time in all species, but in the nearly neutral model, more should fail to be eliminated in species with large bodies (small populations) than in those with small bodies (large populations). The complexity model would probably predict the opposite. Genes in complex organisms may well have to multitask, as we don't have all that many. Remember, we have about 20,000 protein-coding genes, about the same as our millimeter-long lab worm. If our proteins are multitasking to enable cell type diversity, they should be especially slow-evolving and highly constrained. At the least, they should be no faster-evolving in us compared to simpler flies or aphids.

We can repurpose our Ka/Ks calculations to get at this, as Ks − Ka is telling us about the number of mutations that happened and were let through because selection is too weak. We can then normalize that against Ks, so getting a number between 0 and 1, which we can call constraint. Zero means there is no constraint on proteins and the rate at which protein-changing mutations are let though is the same as the background rate of evolution. One means selection is great at removing deleterious mutations that change our proteins.

If the nearly neutral theory is right, for us the constraint on our genes should be low, as selection is inefficient owing to our low population size. Conversely, if the complexity model is right, for us the constraint on our genes should be high, as we have lots of different types of cells.

Ohta did several early tests of this sort of prediction and found that proteins evolve faster when the population size is low, as fits her nearly neutral model. A bigger-scale test, looking not just in mammals but more broadly, found just what was predicted by the nearly neutral model: the constraint operating on our proteins is weaker. More generally, there is a trend for the constraint to be stronger when generation times are shorter (short generations are associated with larger populations)—just as the nearly neutral model predicted (fig. 5.2). As this is the average for lots of proteins in the same genome, and nothing to do with, for example, number of genes or complexity of the DNA, this suggests that, just as the nearly neutral model predicts, selection just isn't all that important when considering the DNA of species with large bodies and small populations. Similar tests found that some of the bits of our DNA that control when a gene switches on or off (so-called promoters) also decay faster when population sizes go down.

The nearly neutral model also makes good predictions about the sorts of changes to proteins that we expect to see. Mutations can change one amino acid to another that is chemically different (big-effect mutations), or to one that is rather more similar (small-effect mutations). When population sizes are small and selection not so great, selection may be able to weed out the really damaging mutations, but not the smaller-effect ones. In bigger populations all such mutations could be expunged from the population. Jianzhi Zhang did such a test, and indeed found that in small populations like ours, selection can remove big-effect mutations but not the small-effect ones. By contrast, in large populations, there was no such difference.

But can we really be confident that complexity doesn't explain the trends claimed by the nearly neutral model? One route to ask about this is to ask about organisms with more or less the same complexity, the same number of different cell types. We can do that in mammals, and again the nearly neutral does well. Intron sizes, for example, are smaller

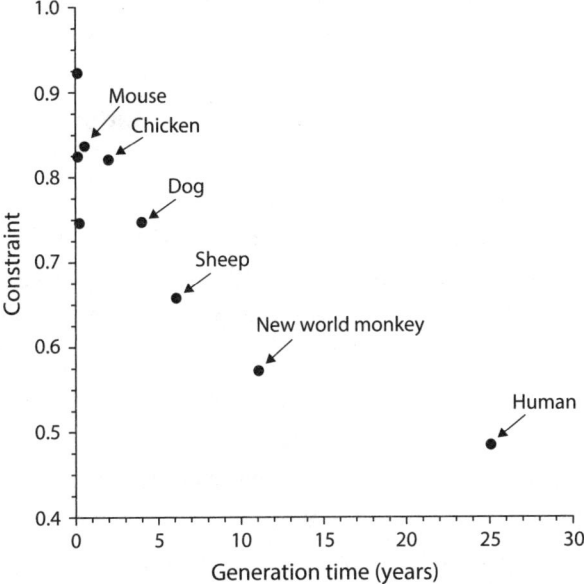

FIGURE 5.2. **The relationship between the estimated fraction of amino acid mutations that are eliminated by selection (constraint) and generation time.** The three unannotated points are all fruit flies. Data from Keightley, P. D., and Eyre-Walker, A. 2000. "Deleterious mutations and the evolution of sex." *Science* 290:331–333.

when the effective population size is higher, even if the organisms have about the same degree of complexity. The introns of mice are indeed shorter than ours. It is not a perfect test, as again, low population size usually means bigger, so it could be owing to something else, such as metabolic rate. Proteins are also slower-evolving in mammals with larger populations than in those with small populations.

Perhaps a better test would be to compare the same species—or recently diverged species—when on an island versus on the mainland. The idea here is that if the island species has a smaller effective population size, selection should be less efficient. It is a great premise, and well controlled for complexity. Unfortunately, the test doesn't work because, for reasons unknown, island species and mainland species have more or less the same effective population sizes. A shame—it was a good idea otherwise.

Overall, then, while we still have no good solution to the complexity problem, the nearly neutral model seems to make risky and correct predictions. In this view, our cells are accident-prone cities and our DNA is mostly nonfunctional junk. We are not simply the product of an eternal March of Progress.

How, you might wonder, do we square this view with the fact that we in other regards seem a rather brilliant species? We live a long time (if we make it past age ten), have protracted senescence, and don't greatly overproduce offspring (unlike salmon and flies). Is there not here a paradox? There is no paradox. Remember that insertion of a jumping gene into a section of unused DNA can easily fly under selection's radar if its effects are minuscule (as commonly we expect them to be). By contrast, if you imagine a mutation that causes changes to the size of offspring and hence the number of offspring, these can very easily be of much larger selective effects. These larger selective effects are not obviously influenced by subtle effects of effective population sizes. Some problems sit within the domain considered by the nearly neutral theory (small-effect mutations); many don't. Small effects on DNA anatomy and behavior are expected to commonly sit in the domain of the nearly neutral theory; mutations that affect organismic anatomy and behavior, much less so.

Wald's Advice for Genomes

If Abraham Wald, sent to decide where to reinforce US bombers as we saw in the last chapter, was advising where to reinforce our genome (protect it from mutations), I'm sure he would strongly advise the protein-coding genes (our engine and cockpit) and not the material that takes mutational hits that seem to be of no consequence. There are some suggestions that organisms might be able to do this, but the jury is still very much out.

He would, I am sure, also have another word of advice: try not to fly through a sky full of bullets.

A rather obvious truth is that the higher the number of bullets/mutations you receive, the lower your chance of survival. This insight underpins work on the problem of the mutation rate. The obvious conclusion is that selection should favor a lower mutation rate, as mutations tend to be damaging. But here again population size intervenes. Not only are we poor at getting rid of bad mutations, inefficient selection also means we generate lots of them. We suffer a double whammy. This is the story of the next chapter. It has great consequences for us, as it tells us why rare genetic diseases are remarkably common.

6

Why Rare Diseases
Aren't Rare

John Maynard Smith, a leading figure in mathematical approaches to the study of evolution in the second half of the twentieth century, was the very epitome of the brilliant, nutty British professor—unkempt white hair, round thick-lensed glasses, an old Etonian with communist proclivities. JMS (as he was affectionately known) was once interviewed for a television series in which luminaries were asked to nominate their "wonders of the world." I recall his answer for all its singular peculiarity: the mutation rate.

The mutation rate that he was thinking about was the rate at which you are born with new mutations, ones that neither your mother nor father had themselves inherited. This is the so-called "germline" mutation rate. The mutations happen in the cell lineage—or line of cellular descent—between the fertilized egg that was your mum or dad and their germ cells (sperm and eggs). Mutations also happen in the cells of our bodies: our livers, legs, or lungs. These so-called *somatic mutations* are not passed on. They can, however, lead to cancers.

We have only recently started to get a handle on somatic mutation rates, so I'll consider them only briefly toward the end. The germline rate we much better understand. This is also the rate that is relevant to the process of evolution. When Maynard Smith was speaking, we had a pretty good idea about what the human (germline) mutation rate is. Now, by sequencing the DNA of parents and their kids, we have a very good understanding of what it is. As a number it is about 1.4×10^{-8}. There remains a bit of disagreement, as it depends on just how you measure it—and on the age of your father. But it is somewhere around there. Written differently, this is 0.000000014. JMS's point was that this number is very small. He thought it was a wonder that it could be that small.

Let's first unpick this number a bit more. What the 1.4×10^{-8} number means is that if you sequence 100 million base pairs of DNA from any of your chromosomes (this being 10^8 base pairs), and sequence the same section from your mum or dad (whichever the DNA came from), then on the average you will find about one new mutation that happened in mum or dad that you inherited. There is a reason we assume that mutations are rare.

For purists out there, you might blanch at the idea that these are called *rates*, as rates usually are measured per unit time: sixty miles per hour is a rate of travel. In most cases (viruses are an occasional exception) we usually measure the mutation rate as a rate per generation, not per year. I'll explain why later.

Semantics aside, I think we can all agree that 0.000000014 is indeed a small number. But it is also rather large. While the human rate is low—in absolute terms—it is one of the highest mutation rates we know (fig. 6.1; table 6.1). To date, the only one higher is of a strange fungus called *Neurospora* that uses mutations as its bullets to target jumping genes. The lowest claimed value is about 1,000 times lower than our rate, reported in a single-celled species, *Paramecium tetraurelia*.

For us at least, it gets worse. There is a second way to calculate the mutation rate and that is to ask how many new mutations you will be born with in each genome, not the number per base pair. Species with more DNA will have more mutations if all else is equal. For us this figure is about 10–100 new mutations per genome per generation, 40–50 on average. For most other species the number is way under 1 (see table 6.1). For lowly yeasts (*Saccharomyces cerevisiae*), for example, the comparable number is about 0.003.

Not a problem, you might think, given that so little of our DNA does anything useful. If we figure this in—let's assume 10% is under selection, as we saw in the last chapter—then we are all born with somewhere

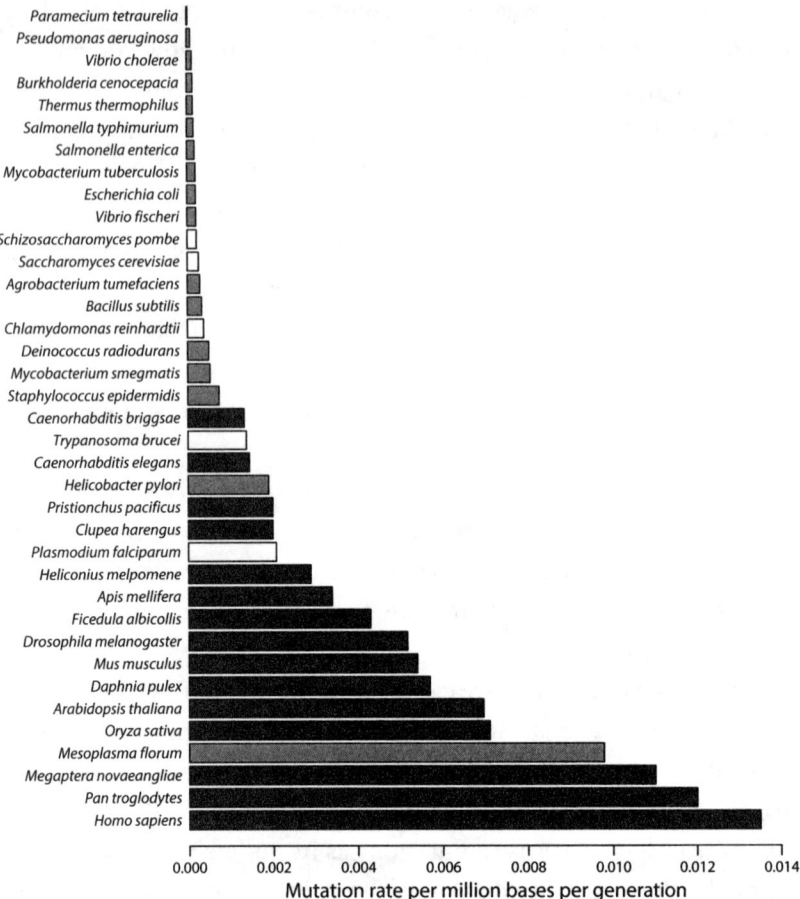

FIGURE 6.1. **Mutation rates per million base pairs per generation for a variety of organisms.** For common names see table 6.1. Black bars are multicellular species; white are single-celled species with, like us, a nucleus; gray are bacteria.

around 5 to 10 harmful mutations per genome (a minimum benchmark is probably around 2). It may have been a little lower earlier in our evolutionary history, as older fathers are the source of more mutations. We are not too sure about the comparable proportion of sequence that is functional for other species, but the total amount of DNA that codes for protein (we refer to coding sequence as CDS) is probably a good proxy. By this measure we still have one of the highest known rates. Our best estimate is about 0.5 new mutations in our protein-coding exons per

Table 6.1. Mutation Rates for a Variety of Species.

Number	Species	Classification	Mutation rate (per bp)	Mutation rate (per genome)	Mutation rate (per CDS)
3	*Homo sapiens*	Human	1.35E-08	44.55	0.49253
11	*Pan troglodytes*	Chimp	1.20E-08	42.288	0.44628
36	*Megaptera novaeangliae*	Humpback whale	1.10E-08	30.14	0.483
25	*Mesoplasma florum*	Bacterium	9.78E-09	0.0077262	0.00718
10	*Oryza sativa*	Rice plant	7.10E-09	2.7619	0.71945
2	*Arabidopsis thaliana*	Plant: laboratory weed	6.95E-09	0.831915	0.29239
5	*Daphnia pulex*	Water flea	5.69E-09	1.4225	0.17165
9	*Mus musculus*	Mouse	5.40E-09	14.6718	0.19154
6	*Drosophila melanogaster*	Fruit fly	5.17E-09	0.872	0.11967
35	*Ficedula albicollis*	Collared flycatcher	4.30E-09	4.8074	0.118
1	*Apis mellifera*	Honey Bee	3.40E-09	0.8908	0.0985
7	*Heliconius melpomene*	Butterfly	2.90E-09	0.793991	0.11339
15	*Plasmodium falciparum*	Malaria parasite: single-celled eukaryote	2.08E-09	0.047528	0.02513
37	*Clupea harengus*	Herring	2.00E-09	1.7	0.102
12	*Pristionchus pacificus*	Nematode worm	2.00E-09	0.3394	0.05932
24	*Helicobacter pylori*	Bacterium	1.90E-09	0.0031445	0.00288
4	*Caenorhabditis elegans*	Nematode worm	1.45E-09	0.145435	0.03625
18	*Trypanosoma brucei*	Sleeping sickness parasite: single-celled eukaryote	1.38E-09	0.0359835	0.01811
3	*Caenorhabditis briggsae*	Nematode worm	1.33E-09	0.13832	0.032
31	*Staphylococcus epidermidis*	Bacterium	7.40E-10	0.0018981	0.00155
26	*Mycobacterium smegmatis*	Bacterium	5.27E-10	0.00368373	0.00343
22	*Deinococcus radiodurans*	Bacterium	4.99E-10	0.00163872	0.00147
13	*Chlamydomonas reinhardtii*	Alga: single-celled eukaryote	3.80E-10	0.04221838	0.01487
20	*Bacillus subtilis*	Bacterium	3.35E-10	0.00143581	0.0012
19	*Agrobacterium tumefaciens*	Bacterium	2.92E-10	0.00165681	0.00146
16	*Saccharomyces cerevisiae*	Baker's yeast: single-celled eukaryote	2.63E-10	0.00327777	0.00229
17	*Schizosaccharomyces pombe*	Fission yeast	2.17E-10	0.00425971	0.00156
34	*Vibrio fischeri*	Bacterium	2.08E-10	0.00088899	0.00077
23	*Escherichia coli*	Bacterium	2.00E-10	0.000928	0.00078
27	*Mycobacterium tuberculosis*	Bacterium	1.95E-10	0.00085898	0.00079
29	*Salmonella enterica*	Bacterium	1.74E-10	0.00084599	7.00E-04

(*continued*)

Table 6.1. (*continued*)

Number	Species	Classification	Mutation rate (per bp)	Mutation rate (per genome)	Mutation rate (per CDS)
30	*Salmonella typhimurium*	Bacterium	1.52E-10	0.00073872	0.00066
32	*Thermus thermophilus*	Bacterium	1.38E-10	0.00029394	0.00029
21	*Burkholderia cenocepacia*	Bacterium	1.33E-10	0.0010245	9.00E-0
33	*Vibrio cholerae*	Bacterium	1.15E-10	0.00045368	4.00E-0
28	*Pseudomonas aeruginosa*	Bacterium	7.92E-11	0.0005171	0.00047
14	*Paramecium tetraurelia*	Ciliate: single-celled eukaryote	1.94E-11	0.00139864	0.0011

Note: Rates are given as per bp per generation, per genome per generation, and per total CDS (i.e., rate of mutation in all protein-coding genes). They are presented in rank order by mutation rate per base pair, high to low.

genome per generation. Rice plant, which has about three times more protein-coding DNA than us (go figure), may have slightly more: 0.7 is their figure (see the right-hand column in table 6.1).

Overall, then, no matter how you measure it, the human mutation rate may be low in Maynard Smith's eyes, but viewed in the round, we have a problem, both literally and conceptually. We have an unusually high mutation rate.

Why then is our mutation rate low—but also high? How, more generally, can we account for the variation between species in their rates? With recent advances in DNA sequencing technology, allowing us to get ever more and better estimates, we seem to be arriving at an answer. Before that, we need to think about how the mutation rate is expected to evolve.

Why Selection (Usually) Favors a Low Mutation Rate

For many, these numbers are a surprise on two fronts. First, mechanically these represent amazing acts of copying. Many mutations happen during the process of copying DNA during cell division. Occasionally the non-complementary base pair is added—i.e., where it should not go. There could be an A on one strand and in copying the DNA we erroneously added a G, perhaps, instead of a T. We now have an AG mismatch in the DNA, not an AT as nature would intend. If this is repaired by patching in the G with a C, a mutation will have happened. That the mutation rates are so low indicates that this sort of mistake is rather rare.

The second reason is that it is tempting to adopt a line of reasoning that is interestingly (nearly always) wrong. It usually runs like this: evolution requires variation, mutation provides variation, therefore evolution should favor a high mutation rate. It looks logical, doesn't it? Why then is it wrong?

One reason is that evolution is not a thing requiring anything. It is a name we give to a process. Perhaps we could be more precise: for species to survive in the long term, the individuals in that species need to be able to adapt to their changing environments; those with higher mutation rates will be better able to adapt; therefore selection should favor a high mutation rate. It still sort of looks logical, doesn't it? What could be wrong?

It may be the case that some species have properties that make them more, what shall we say, evolvable. The problem comes in the last clause: that because evolvable species do better in the long term, this explains why they have the properties that make them evolvable, such as a high mutation rate.

As we saw in chapter 3, we now know that there is likely to be an error here: today's selection cannot favor something that will be useful in the distant future, especially if it is currently harmful. It couldn't preserve the ability to synthesize vitamin C millions of years ago just because modern-day sailors would now find it beneficial. Selection cannot also favor a high mutation rate now just because an asteroid might hit sometime in the future (FYI, an asteroid probably explains the extinction of dinosaurs, as the impact changed the global climate). It is perfectly reasonable to ask why some sorts of species persist through such change more than others. In the case of the extinction event that killed the dinosaurs (except birds, which are flying dinosaurs), the victim species, as they are known, tended to be those that lived in trees. But you can't then leap to the conclusion that survivors were ground-dwelling prior to the asteroid because selection favored this on the basis that, should an asteroid hit sometime in the future, it would be good to now be ground-dwelling.

As before, the question we should instead ask is: What would happen to a new mutation that itself changes the mutation rate in the here and now? Organisms have repair enzymes that, for example, attempt to check whether there have been non-complementary base pairings or missing nucleotides on one strand, and if there are, to correct them. These explain in part why mutation rates are so low—the errors happen, but then are patched up. That AG mismatch can be repaired back to AT before any real harm is done, a bit like a bullet hole in the side of a plane being immediately filled in. A mutation in the gene coding for such a repair enzyme, by altering the enzyme, can decrease or increase the mutation rate (usually increase) for all the DNA in a genome. We see these in bacteria and yeast: some sub-strains have higher mutation rates owing to repair enzyme mutations. Recently, it has been discovered that there is variation between mice in their germline mutation rate owing to variation in the gene for a repair enzyme, *Mutyh*. In humans, mutations in this gene predispose to colon cancer. In humans, hypermutators seem to be rare. Analysis of over 21,000 families with genetic diseases found just two incidences of a mutation in a repair enzyme associated with elevated germline mutation rates.

What is going to happen to a mutation that alters the gene for a repair enzyme and increases the mutation rate? Remember, we are trying to ask whether selection should favor future evolvability, so a higher mutation rate. It is true that without variation there can be no evolution, but most of the time selection will not persistently favor an increase in the mutation rate. This is for the simple reason that most mutations are harmful. There may be the occasional one that is advantageous—especially when a species is faced with a new threat or a new environment—but mostly they are just bad for you. That is what Ka/Ks <1—true for most genes—tells us. A mutation increasing the mutation rate may create an advantageous mutation or two—but this will usually be swamped by the increase in the number of harmful mutations.

You can do the experiment in bacteria. Give some a high mutation rate, some a lower one, and introduce the high-mutation-rate ones at low frequency into the population of low-mutation-rate bacteria. The high-mutation-rate ones usually go down even further in frequency. An alternative experiment is to give different bacterial strains different mutation rates and see how well they adapt to novel environments. Such experiments have revealed that the highest mutation rates were associated with an *in*ability to adapt to new conditions (so much for evolvability) and that the surviving high-mutation-rate bacterial strains had instead evolved lower rates of mutation. Mutation is for the most part not a good thing for bacteria or humans. Indeed, mutations in repair enzymes also underpin nearly all human genetic diseases associated with accelerated aging, such as Werner syndrome.

A second reason is more subtle. In a sexual species the odds are doubly against a mutation that increases the mutation rate. Individuals with a new not-so-good repair enzyme have higher mutations rates in all genes. Most of these mutations are harmful. But occasionally there will be a beneficial one, like the mutation that converts white moths to darker ones, which, if trees are dark, would be beneficial. Let's suppose this has happened. Perhaps the beneficial mutation (e.g., the dark-pigment mutation) could now increase in frequency owing to its selective advantage. But what happens to the mutation that increased the mutation rate? Notice that we are now talking about two different genes, each with a mutation: a repair enzyme gene and a color gene. Each has two different versions: high/low mutation rate, white/dark color.

The problem sexual species have is that these two genes are likely to be on different chromosomes. Consequently, each sexual generation both "flavors" of the two genes are transmitted independently when they go to eggs and sperm. The consequence of this is that, while the black color mutation goes up in frequency, there is no reason the high-mutation-rate version of our repair enzyme would too. As they are on different chromosomes, their fates are uncoupled.

An assumption here is that the not-so-good repair enzyme is itself neither directly beneficial nor harmful. Its fate is then dependent on chance (drift/bobbling about) and, crucially, the genomic company that it keeps. As they are on different chromosomes, there is no reason the high-mutation-rate mutation can stay in the same individuals as those with the new helpful condition (being black if the background is black). As, more generally, individuals with the high-mutation-rate repair enzyme have more harmful mutations, on the average it surrounds itself in a sort of genetic cloud of bad DNA. It keeps bad genetic

company. On the average it will be in a less-well-off-than-average individual and so will be eliminated from the population. The same recombination process—chromosomes going their own ways independently
during reproduction—allows the beneficial mutation to escape this
toxic genomic environment.

In broad outline, then, we think we know why mutation rate numbers
are so small. Were I to ask you to choose between flying a plane through
a sky riddled with bullets or one largely bullet-free, you would be somewhat deranged if you chose the former. Selecting the former is wishful
thinking. You might suppose that in the volley of bullets will be one that
makes a change for the better to the plane. You probably would not
survive to discover whether you were right.

Is it then impossible to select for a higher genomic rate of mutation?
Probably not. In half of Lenski's twelve bacterial lines in his long-term
evolution experiment, the mutation rate went up, so increased rates
when a population is adapting to a new environment are possible. We
can also, for example, experimentally force conditions where the need
to diversify outstrips the harm resultant from increasing the mutation
rate. Given the above logic, we would need to do this in a species that
can't recombine, can't shuffle its DNA. A bacterium with one chromosome would be a possibility. Second, we would need to force conditions
that strongly favor continuous short-term change.

One such experiment took replicate populations of bacteria and into
one introduced a bacterial virus (known as a *phage*—yes, bacteria get
viral infections as well). The other was phage-free. The researchers then
kept the infected population playing cat and mouse with the phage over
many generations. In some of their pairs of experiments (with phage
versus without phage), the ones with the infection evolved higher mutation rates. We presume that the inability to escape from this virus (imposed by the experiment), and possibly reduced variation over time in
the closed confines of the experiment, meant that selection could sometimes favor an increased mutation rate, as occasionally a mutation useful
in the cat-and-mouse game would crop up. This useful mutation would
spread in the population. As bacteria are not sexual, and have only one
chromosome, the spread of this beneficial mutation (we don't know

what it might have been) drags with it the version of the repair enzyme that wasn't that great. Incidentally, this process whereby the increase in frequency of a good mutation causes those in the same genome (usually on the same chromosome) to also increase in frequency has a name: hitchhiking. John Maynard Smith came up with that term.

There may also be conditions that temporarily favor a programmed increase in the mutation rate. When bacteria get DNA damage, or are just stressed, they start a so-called SOS response that, among other things, increases the mutation rate genome-wide. This may be beneficial—if you are so damaged that you are about to die, then why not mutate in the hope of finding a solution? But it might not be—if I am damaged, perhaps I will suffer a raised mutation rate as collateral to my attempts to repair myself. Recent evidence supports the latter. In *Pseudomonas* bacteria exposed to low levels of an antibiotic, this being the stressor, the SOS response is generally helpful in coping with the stress of being exposed to the antibiotic, but the increased mutation rate, which is modest, is not a component of that. It appears to be a stress-induced unwanted side effect.

Why Doesn't Everything Have the Same Mutation Rate?

Selection for increased mutation rates, we then think, is a second-order effect. The primary effect is to reduce the rate, as most mutations are harmful. Mutation rate theory is, in some regards, odd, as at first sight the prediction is that everything should be the same and (just about) everything should have no mutations. If you think of mutations as unwanted cellular accidents, you can see why. But as we have seen, while numbers are very low, there is lots of variation. As noted above, *Paramecium tetraurelia*'s mutation rate per base pair is 1,000 times lower than ours. We have 150,000 times more mutations per genome than *Thermus thermophilus*, a bacterium that lives near thermal vents and thrives at about 65°C (145°F). This equates to nearly 2,000 times the number of mutations in protein-coding genes. What can explain this variety?

For many years now there have been a series of ideas swirling around to explain the variation in mutation rates. It may well be the case that they all have some relevance. Nonetheless, as we shall see, the nearly neutral model seems to be the most important of these.

When we talk about the mutation rate per generation, this means something different for single-cell species than it does for multicellular ones like us. For the latter the "per generation" refers to the per sexual generation. In effect, our mutation rate is equivalent to the number of mutations from fertilized egg (your parents when they were each a single cell) to fertilized egg (you, when you were just one cell). These mutations happened at some time in the cell-lineage history of the sperm and egg in your parents. In this process, there are many cell divisions. In women, if we track back from any egg cell in a woman's ovary to the initial fertilized cell (the zygote) that became the woman, we think there were about 30 cell divisions in between. In men, especially older men, there are quite possibly many more than that between any sperm a man produces and the fertilized egg that became that man. This is because in men, sperm are made from a sperm stem cell that keeps dividing— one of the daughter cells becomes a sperm, the other stays as a stem cell, capable of doing the same process repeatedly. The older the man, the more such divisions, the more chance for errors during cell division.

In single-celled species the mutation rate is defined per cell division. In yeast, for example, its single cells divide all the time. Here we consider one of the daughter cells and compare it, in principle, to the parent cell.

One problem is that this doesn't look like a fair comparison. How could it ever be appropriate to label both measures—per cell division and per sexual generation—as "per generation"? Notice, for example, that the multicellular species have higher per-generation mutation rates (fig. 6.2). Could this not be a trivial consequence of having many more cell divisions contributing to each "per-generation" measure?

If you thought this, then you would have a point. However, the reason we do it this way comes back to the problem of selection on the mutation rate. To a first approximation, in a sexual species like us, the mutations that matter are those that affect how well we perform *as individuals*. Put differently, the selective fate of a mutation that makes a moth black is

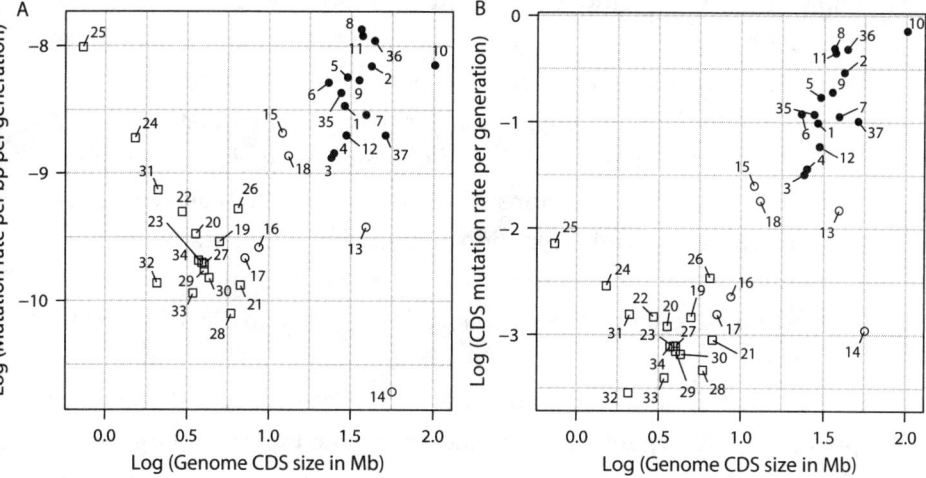

FIGURE 6.2. **How mutation rates vary with the amount of protein-coding sequence in a genome. a.** Log of the mutation rate per bp per generation. **b.** Log of the total number of mutations in coding sequence per generation. Circles are eukaryotes (have a nucleus in the cell), squares are prokaryotes (i.e., bacteria). Open figures are unicellular species, filled data points are multicellular. Numbers relate to species numbers in table 6.1.

largely owing to its effects on moths, not its effects on a moth's sperm or eggs. If that is generally true, then the mutation rate we need to worry about is the one that impacts between-individual selection, not events in the cells giving rise to sperm and eggs. For single-cell species, each single cell has its own independent fate; thus the relevant measure is the mutation rate per cell division. Selection on a mutation that increases the mutation rate would be mediated by mortality at the cell level for single-celled organisms, but at the individual level in multicellular species.

This is not to say that there cannot be selection that operates on different cells within individuals—cells can indeed compete to become the sperm and eggs in our bodies. Human females start out with 6,000,000 potential eggs when they are babies, but in their lifetimes they use only a few hundred, at most. We think there may well be some sort of selection weeding out good from bad potential eggs. In males, too, we know of certain mutations that seem to predispose pre-sperm cells to become the sperm cells by outcompeting other potential pre-sperm cells. In several cases we think these mutations that are good for pre-sperm are also bad for people. A fingerprint of such mutations is that nearly all of the mutations in affected kids are inherited from the father. The human genetic conditions Apert syndrome, Noonan syndrome, achondroplasia, and multiple endocrine neoplasia 2B are all thought to be manifestations of selection favoring mutations when good in the competition to become a sperm, but bad for us, the end users of the same mutation.

Leaving this issue aside, for many years it was thought that the key predictor of the variation in the mutation rate was the size of the genome, or rather, the total amount of functional sequence. This makes sense, doesn't it? Selection cannot really care about the per base pair mutation rate as such; it cares about the number of new harmful mutations that you are born with and the net harm that they do. If you have more protein-coding genes—or more functional DNA more generally—then what matters is how many of these end up badly functioning because of mutation.

Going back to the plane analogy, if you have a plane that is big enough to need four engines, you are going to spend more money on making all four bulletproof than if you only have two engines. There is a weakness in this analogy, however. A plane can fly with three of four engines. When thinking about genes we are thinking more about genes that are not doing the same job. We would need all four such genes to be reinforced to the same degree, as they are all vital.

Early data were supportive of this idea. Looking at the best estimates at the time for various single-cell critters (bacteria, fungi, etc.) and some viruses, it was observed that the genomic mutation rate was approximately constant across all the species (about 0.001 to 0.01 mutations per genome per cell division). As the genome size varied greatly among these species, the per-base-pair rate was much higher when genomes were small. This looks like strong evidence that genome size is the big determinant of the variation in the per-base-pair mutation rate: assuming they have absolutely more functional sequence, bigger genomes need lower mutation rates (per bp), more functional DNA needing to be protected from mutation.

There are now two issues with this idea. The first is that, with more (and better) data, it presently isn't quite so clear that it applies across single-celled species, but it may well. With more data we will know. The second problem is that this trend reverses when we add in multicellular species. Even allowing for how much of their DNA codes for proteins, the larger the genome (or more accurately the more DNA coding for protein) the *higher* the mutation rate per base pair (fig. 6.2a). This is the opposite of the original finding. The total number of mutations in genes also goes up with the number of genes (fig. 6.2b). This is also not what was originally seen: it was originally thought that the total number of mutations was constant, no matter the size of the genome.

Perhaps then there is no single theory of the mutation rate? Perhaps we need different theories for different organisms? Perhaps we need to go back and think about whether we should be measuring everything per cell division. Well, let's not be quite so hasty. We haven't yet heard from the nearly neutral theory. What would it predict?

Remember that the nearly neutral theory says that selection is just much less efficient when population sizes are small. Could this also apply to mutations that alter the mutation rate? As Mike Lynch has forcefully argued, yes, it could. Selection should (nearly) always favor a low mutation rate. That the numbers are usually so very low seems to support this. As the mutation rate gets ever lower, a mutation that weakly increases the mutation rate will not be a deleterious mutation but will, in Ohta's framework, be a weakly deleterious or an effectively neutral mutation. There will be a balancing point between drift and selection. This is the *drift-balance* hypothesis, a corollary of the nearly neutral model.

Core, then, to the drift-balance model is the prediction that mutation rate of any species should vary with population size. In large populations selection should be an efficient force—just as it keeps rates of gene-processing errors down, so too it could keep mutational errors down. Mutations are, after all, mistakes, errors. In species like us with a low effective population size, by contrast, the balance point will be at a higher mutation rate.

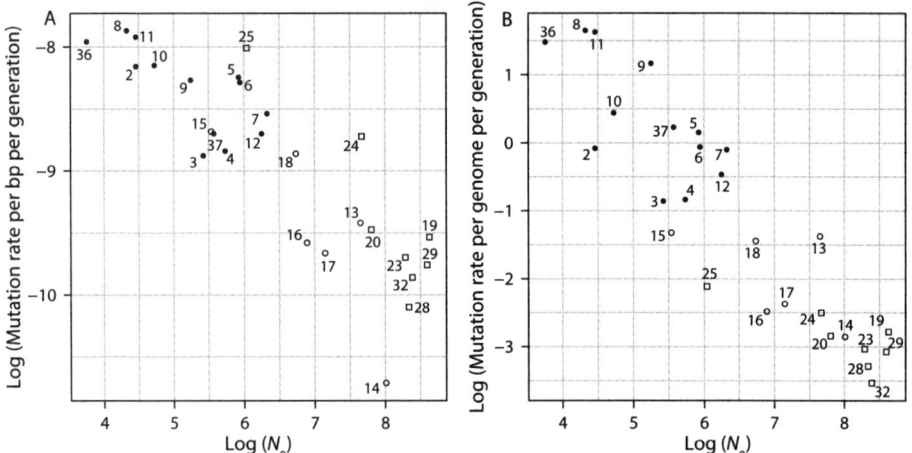

FIGURE 6.3. **Mutation rate as a function of effective population size** (N_e). **a.** Log of mutation rate per bp per generation. **b.** Genomic mutation rate per generation. Numbers relate to species numbers in table 6.1. Circles are eukaryotes (have a nucleus in the cell), squares are prokaryotes (i.e., bacteria). Open figures are unicellular species, filled data points are multicellular.

The theory thus predicts that the smaller the population, the higher the mutation rate. This is not because small populations have a great advantage from having a high mutation rate—they don't. It is just because, like all errors, they aren't very good at stopping them from happening. We read through stop codons and we mutate too much. And there isn't much we can do about either.

Is there any evidence for this? With recent advances in DNA sequencing, we can now put together lots of estimates of both variables—effective population size and the mutation rate—and it seems to be a good fit between prediction and evidence (fig. 6.3). The per-base-pair mutation rate is lower when the effective population size is higher (fig. 6.3a) and the total number of mutations is higher when the population size is lower.

This looks like what some might call a slam dunk. And I would be tempted to agree. There are, however, a few concerns. First, I didn't tell you how we measure the effective population size. Presently we use a combination of the mutation rate and the amount of genetic variation within a species. The idea, derived from the neutral theory, is that species with larger populations will have more mutations just bobbling about (i.e., more variation within a species), but that this number is also affected by the mutation rate. The higher the mutation rate, the higher the rate at which we expect to see mutations just bobbling about. This creates a little concern, as we are using a number (the effective population size) to predict the mutation rate that itself depends on the mutation rate. However, this isn't a problem if we can measure things accurately enough, and we think we probably can. Recent evidence indeed supports this, and further suggests that longer generation times (associated with small population sizes) and larger genomes (also associated with small population sizes) are predictive of higher mutation rates. It should also be emphasized that there is quite a bit of scatter in the data. Indeed, the humpback whale has the smallest population size in the above sample but not the very lowest mutation rate. It might just be that being massive imposes further selection on the germline mutation rate, for example to reduce cancer risk, as the single cell of the fertilized egg must go through many more rounds of cell division to make the whale's body.

A second concern is that while the trend is as expected, it doesn't follow that organisms are actually at the mutation rate predicted by the drift-balance model. An alternative is that organisms have a mutation rate above this position and that the mutation rate is also determined by, perhaps, the costs of keeping the rate down. Larger genomes or more cell divisions in the germline might both increase the costs. An alternative possibility is that, while selection typically doesn't favor very high mutation rates to enable evolvability, there could at least in theory be an adaptability advantage to a slightly raised mutation rate if the environment changes often enough. There is some evidence that real-world mutation rates can be a little higher than predicted by the drift-balance model. In yeast, for example, if you relieve them from selection in the lab, many times they will evolve a reduced mutation rate. This suggests that the mutation rate seen in nature could have been lower than predicted by the drift-balance model and that, instead, there is some sort of balance between selection reducing the mutation rate and costs of keeping it low, which balances out differently depending on the effective population size.

Our Mutational Double Whammy

It seems then that we have a good understanding of why some species have a low mutation rate and others have a higher rate: it has in no small part to do with the ineffectiveness of selection when populations are small. But note, too, the species with the smallest effective population size and the highest mutation rate in our sample, no matter whether it is measured per base pair or per genome. It is us.

In effect, selection (or the lack thereof) kicks us when we are already down. With a low population size, our DNA decays and gets filled with junk, just because selection cannot do anything about it. But on top of that, the rate at which harmful mutations come in is also relatively high, again just because selection cannot do too much about that, either. A double whammy.

One immediate consequence of this is that we have a high rate of genetic diseases. Rare diseases are typically the result of mutation and are so named because few individuals suffer from any given disease. There is no agreed definition of how rare a rare disease needs to be to be considered rare. In the US, according to the Orphan Drug Act, any disease affecting fewer than 200,000 people is "rare." This is less than about one in every 1,650 people. The EU defines a rare disease as one affecting less than 1 in 2,000. The Koreans have the most restrictive definition, with a prevalence of less than 1 in 20,000. Regardless of the definition, about three-quarters of those affected are children, and about a third of those children will not make it to their fifth birthday. Four-fifths are genetic conditions.

We think we understand why each rare genetic disease is rare (usually much less than one in 2,000). For any disease that kills us early in our development, its frequency in the population is a balance between the rate at which new mutations arrive in the population and their loss owing to this death. In species other than humans, this "loss owing to death" is otherwise known as purifying selection. In humans it is a raw and personal family tragedy. It often seems to me wrong to discuss such deaths in cold evolutionary terms. Each "selective mortality" is so much more than a number in our equations.

Whether in humans or not, the balance point between mutations coming into a population at some rate and leaving at an equal but opposite rate is known as *mutation-selection equilibrium*. We can calculate what sort of frequencies we might expect to see—we predict low frequencies of each rare disease. The balance point is dependent on the chance that the mutation will kill you; whether it is recessive (you need two mutant versions to get the disease) or dominant (one mutation is enough); and the rate of mutation of the gene that results in the disease. The higher the mutation rate, the more common it will be. The best predictor of this is the size of the protein encoded. Larger genes for larger proteins will have more mutations, just because they are large. The genes for breast cancer (*BRCA1* and *BRCA2*) and for Duchenne muscular dystrophy (*DMD*), for example, are some of the largest genes we have. DMD, the disease, usually only affects boys, and the frequency of births is about one in 4,000 male births. Inherited *BRCA1* mutations account for about 5% of all breast cancers in women under forty. About one in 1,500–2,000 women are born with the mutation, some because it is a new mutation, some because at least one of their parents had also inherited it. Rare, but not that rare. Being a large gene/protein and thus having a high genic mutation rate causes more disease. It also means we study these conditions more.

But rare diseases collectively aren't rare. We have identified over 7,000 different rare diseases, and it is estimated that about 30 million people in the US and 30 million in Europe currently have a rare disease. Those numbers are striking: in the US that is one in every 11 people that has a rare disease. A more conservative estimate is one in 17. Between 5% and 10% seems to be a safe estimate. If extrapolated to the world, this would be about 500,000,000 people currently bearing a rare disease. The numbers might be a bit lower than that, as without advanced healthcare many die earlier than they otherwise would.

The compendium of afflictions both is heartbreaking and, at least to me as a geneticist, holds a fascination. These diseases tell us a lot about what different genes do. You can find online databases of such diseases (try, e.g., rarediseases.org or orpha.net). They make for unhappy reading: Aarskog syndrome: symptoms include short stature and multiple facial, limb, and genital abnormalities. This is associated with mutations in the gene *FGD1*. Since first described there have been about 60 reports. Abetalipoproteinemia, also known as Bassen-Kornzweig syndrome, is associated with inability to absorb fats, resulting in multiple vitamin deficiencies and progressive neurological deterioration, difficulty walking, and blindness. This is associated with mutations in the gene for microsomal triglyceride transfer protein (*MTTP*). Prevalence is unknown but thought to be less than one in a million people. Ablepharon-macrostomia syndrome is associated with the absence of eyelids and a wide mouth. It is caused by mutations in the *TWIST2* gene. Since first described there have been 16 documented cases.

And we aren't even up to the entries starting "Ac."

This tragic scale of suffering can be laid at the feet of our high mutation rate, which in turn can be chalked up to inefficient selection in small populations. If ever you thought that humans were some pinnacle of evolution, these awful numbers must give pause.

You might reasonably ask whether our rate of rare diseases is higher than seen in other species. Recall that of 5,000 salmon eggs, 4,998 will not make it to reproduce. What is their rate of genetic diseases? To some extent this is a meaningless question, as a human genetic disease is defined as a condition bad enough to have a person present to a doctor. A few domesticated species aside, the rest of nature has no doctor to turn to. There are also technical issues. In humans we only talk about "diseases" for conditions that affect you after your birth. Mutations that kill the youngest embryos do not cause genetic diseases, just genetic deaths. The best we can then do is ask about the rates of harmful mutations which can make for a fair comparison between species. As we have seen, in these rankings we are at the top of the charts, or at least close to the top (table 6.1). As we shall also see in the next chapter, when it comes to errors in handling chromosomes (sometimes called chromosomal mutations), we also have the unenviable position of topping the charts. For a species that produces few offspring, it is striking that 60% never it make it to birth, and of those that do, about 10% have a rare disease. While other species must have higher rates of all-cause mortality (including being eaten or failing to eat), we are unusual in having so much death resulting from the DNA bequeathed to us by our parents.

Strange Genetics for Common Diseases

The commonness of human genetic diseases cannot, however, all be laid at the feet of our high mutation rate. Some genetic diseases, like sickle cell anemia and cystic fibrosis, are unexpectedly common. One of these might also challenge the assumption of the randomness of the mutational process, with some evidence suggesting higher rates of adaptive mutations in the contexts when such mutations would be beneficial (but as I'll argue, probably doesn't).

We have assumed that a new mutation could be hidden from view (recessive) or immediately visible (dominant) when it first appears in a population in an individual with one old and one new (mutated) sequence. But there is, at least as regards the fitness of the bearers of one old, one new version of a gene, a third possibility: the mixed type is the best combination.

This might seem odd, but it can account for some unusually common genetic diseases. The best understood is sickle cell anemia. In the United States, sickle cell disease affects about 1 in 365 African American children. In Uganda, possibly the most affected country, just under 1% of the population have the genetic disease (far too high to be considered a rare disease).

We know the underlying genetic cause. Hemoglobin is the molecule that makes blood red (and is found in red blood cells, strange to say). It has two jobs: it is a delivery truck for oxygen (breathed in), and having made the delivery to your tissues, picks up the waste product of using oxygen, carbon dioxide, which then gets breathed out. It is made up of four sub-proteins, two alpha globins and two beta globins. Those with sickle cell anemia have a mutation toward the end of the beta globin gene. A GAG codon is now GTG. Where there should be a water-loving amino acid, glutamate, there is now a water-avoiding amino acid, valine. The mutant version of the gene is known as HbS, as opposed to HbA, the normal glutamate-specifying version.

When there isn't much oxygen around, the beta globin changes shape and exposes the site with (or without) the mutation. With water-loving glutamate, this is of no consequence. For the protein of HbS, it is a problem. As valine repels water, the HbS proteins instead join together—they polymerize. You can think of this as being a bit like you having something sticky stuck under your arms, like chewing gum. With your arms by your side all is well. When you raise your arms, the sticky gum is now exposed and you glue yourself to other people. HbA has no chewing gum under its arms.

In individuals with one copy of HbS and one of HbA (so-called carriers), this isn't much of an issue, except under extreme conditions, such as dehydration or mountain climbing. These carriers are much more common than those with the disease (about 43 million globally, as opposed to 4.4 million with the disease, who have two copies of HbS). For these individuals with one of each, there is normally enough of the non-sticky HbA beta globin to cope. But if you only have HbS, then, when oxygen concentration is low, you have a problem. The long chains of polymerized hemoglobin distort the shape of the red blood cell. What was a smooth doughnut-like structure becomes sickle-shaped (hence the name) and prone to breaking. You end up with anemia (not enough red blood cells), and thus an inability to transport oxygen around the body.

Problems with sickle cell anemia start soon after birth (5 to 6 months). In low-income countries most die before age 5. Even with good healthcare, life expectancy is only about 50 years. Before that, sufferers usually have numerous health problems, as well as both sudden attacks of pain (known as sickle cell crisis) and longer-term pain. Sufferers are also more prone to infections and stroke.

The mystery of sickle cell anemia is why it is so common. Sickle cell anemia is far too common to be explained by the balance between mutation pushing genetic diseases into the population and selection kicking them out. The mutation rate isn't that high! In some countries the HbS version of the gene is about 15% of all copies. That is extremely common for a mutation that, in two copies, kills you before your fifth birthday.

A big clue comes from the global distribution of the disease: about 80% of people with the disease are in sub-Saharan Africa. This also explains why, within the US, for example, it is more common in people with a line of descent from this region. But what is it about sub-Saharan Africa?

Another big clue was discovered by Anthony Allison from the Clinical Pathology laboratory at Oxford's hospital, the Radcliffe Infirmary. In 1949 during a university expedition, he discovered that carriers of sickle cell trait (HbS/HbA) were especially common in populations living near the coast of Kenya and Lake Victoria. In the intervening highlands, these carriers (also technically known as heterozygotes) were much less common. The difference between these two areas, he noticed, was the prevalence of malaria, caused by the single-celled organism *Plasmodium falciparum*: the carriers were more common where malaria was more prevalent. Could it be, he wondered, that the carriers—that had both HbS and HbA—might be less affected by malaria than those with only HbA? Those with HbS alone have sickle cell anemia, so, if he was right, the carriers would be the best off.

To test his idea, he had to wait four years until his return in 1953. As malaria kills the young most, he worked with children from 4 months to 4 years of age. And he found just what he predicted. The children with both the HbS and HbA versions of the gene had some resistance to (or tolerance of) malaria. Not only that, but where this malaria was prevalent across Africa was where the HbS version of the gene was also most prevalent.

Since then, his finding has been confirmed many times over. The potentially lethal complications of malaria, such as when it affects the brain or causes severe anemia, are rare in the HbA/HbS (AS for short) carrier individuals. Further follow-up studies indeed show that, among over 1,000 Kenyan children living near Lake Victoria in 2002, survival of carrier (AS) children exceeded that of both AA and SS children (fig. 6.4). The former died of malaria, the latter of sickle cell anemia. This is the unforgiving process of natural selection in action.

Exactly why—in mechanistic terms—carriers have better resistance to malaria is still not fully clear. However, an enzyme (HO-1) that generates toxic carbon monoxide is switched on in individuals with the HbS mutation. This protects individuals from the worst complication of malaria, cerebral malaria. It seems the HbS *allele* (as different versions of the same gene are called) doesn't affect the life cycle of the parasite in the red blood cells; it just allows AS individuals to tolerate it better.

Evolutionarily speaking, that the carriers survive best can explain the enigma of why sickle cell anemia—a lethal disease—is itself so common. In a region where malaria is common, the mutation to HbS is advantageous, as carriers tolerate malaria better than everyone else who

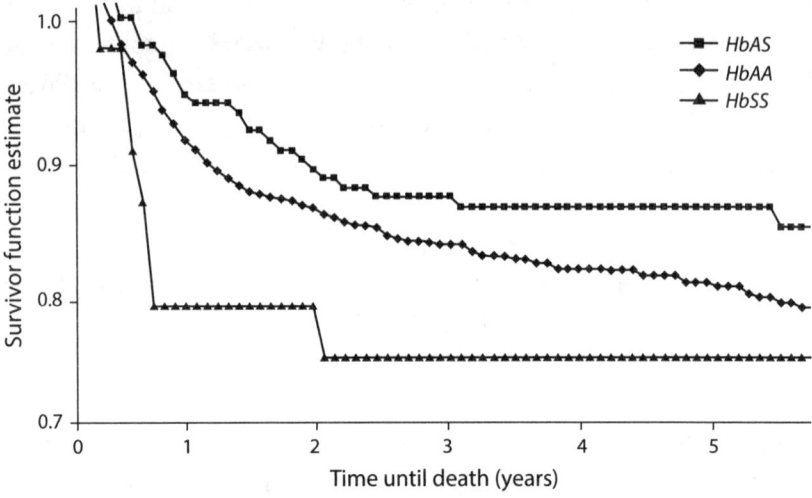

FIGURE 6.4. **Survival of Kenyan children dependent on their Hb genetics. HbAS are the carriers.** Adapted from Allison, A. C. 2009. "Genetic control of resistance to human malaria." *Current Opinion in Immunology* 21:499–505.

is HbA. It increases in frequency because of selection, just as happens with agouti mice, peppered moths, etc. The difference comes when HbS starts to be common. When this happens, the carriers must start inter-mating, and many of their children die, some because of malaria, some because of sickle cell anemia. If two carriers have kids, on average a quarter will be HbA/HbA and vulnerable to malaria, a quarter will get sickle cell anemia (HbS/HbS), and a half will be HbS/HbA carriers. Carriers cannot "breed true" (produce only carriers like themselves) owing to the nature of inheritance. This establishes a balance whereby both forms of the Hb gene are maintained in a population (and we generally call such selection "balancing").

Sickle Cell Anemia Complicates
the Perfectionist View of Nature

In bigger view, examples like this also question our understanding of the evolutionary process, as a new mutation could do well when initially rare, but not as it gets more common. This is a complication for the simpler view of adaptation we saw in chapter 3. It puts a brake on the evolutionary process. It is also a brake to evolving and sustaining perfection. If you ask why everyone can't tolerate malaria, not just the HbS/HbA individuals, the answer is that the inheritance mechanism prevents this. The carriers don't breed true, just because of the way sperm and eggs are made and get together. This exposes an underlying assumption about how selection normally works: normally, carriers cannot be the best type.

What is less clear is how important this issue is to our grand view of how natural selection might shape species. One oddity is that whenever we want to discuss cases where the carriers have the greatest benefit, we always relate the same case history: sickle cell anemia. There seem to be few other cases. There are suggestions that carriers of the cystic fibrosis mutation in Europe must have had some advantage, possibly enabling resistance to tuberculosis, possibly as a defense against diarrhea when milk drinking became common in Europe. But we don't know.

We know that balancing selection isn't an inevitability in response to malaria. In selective response to a different malaria parasite (*Plasmodium vivax*), a mutation in the so-called Duffy antigen gene is now at 100% prevalence through most of sub-Saharan Africa. The Duffy antigen is a protein that sits in the membranous surrounding of our red blood cells and is the port of entry for the parasite. The mutation stops this entry. In this case we see only plain ordinary natural selection, a mutation being beneficial both when rare and when common.

The other reason that carrier advantage (more technically called *heterozygous advantage* or overdominance) might not be too important as a block to evolution is that the effect can be temporary. In the case of sickle cell anemia, the HbS allele is sustained in the population if there is malaria. If the population, for example, moves to a place without malaria, then this selection goes away, and the population heads back to everyone having HbA alone. There is evidence for this from the African American population, where sickle cell anemia is much rarer than in related African populations. However, there is also prenatal screening (a different form of selection), so a reduced frequency is expected because of this as well, among other reasons.

HbS and the Randomness of Mutation

While the case of sickle cell anemia provides evidence that not all human genetic disease can be simply laid at the feet of our high mutation rate, the story has also taken a recent curious turn. It has been presented as evidence that mutation may not be the perfectly random process that we think it usually is. Does this also alter our view of the evolutionary process?

In the canonical view of evolution, mutation throws out a random cloud of mutations, a few good, others harmless or harmful. Selection and drift then operate on these. The mutational process is not the process that gives evolution its direction. That is selection's job.

Repeatedly, however, we see claims that this view is not right, either not quite right (just a tiny bit oversimplified) or, more hyperbolically, that the walls of Darwinian Jericho must come tumbling down, as mutation is a directed process. In this subject area, beware of clickbait.

At the heart of the diversity of possible modes of directed mutation is the notion that mutation rates can be different for different genes or parts of genomes, and that this variation is somehow adaptive. In addition, people sometimes then invoke the idea that the mutation rate is increased in the relevant genes when needed, or that the mutational process is skewed to the creation of mutations that are adaptive.

There are two good examples of something that you might want to call directional mutation in multicellular species—one germline, one somatic. The germline case I alluded to in the introduction. The fungus *Neurospora* has a system known as repeat-induced point mutation, or RIP for short. In the species, to undergo sexual reproduction, a cell forms with two nuclei (one from fungal mum and one from fungal dad). These will then fuse. Just before fusion, in each nucleus they somehow scan their chromosomes for DNA sections that resemble each other at the sequence level—so-called repeats or duplicates. Our genome is rich with such repeated elements, mostly the leftovers of inserted viruses and jumping genes (transposable elements). If *Neurospora* finds such a pair of similar sequences, both sequences then get mutated at a high rate. Hence the name of the process, repeat-induced point mutation.

This, we think, is probably a DNA-level defense against jumping genes—if ever one arrives in the DNA, then hops (by copying itself) from A to B in the genome, it will now be seen as a repeat (A will look like B). *Neurospora* sends mutational bullets to destroy both. Odd to say, the *Neurospora* genome doesn't have many identical DNA repeats. The process is "directed" in the sense that it targets repeats. It is not, however, directed to create adaptive mutations; it is directed to destroy unwanted DNA. In this sense it is no challenge to any conventional understanding of how adaptation works. With Long Wang and Sihai Yang we attempted to estimate the overall mutation rate in *Neurospora* with RIP. It was very high, around 3×10^{-6} per bp per generation, around 100 times our rate. However, the mutations were nearly all in the decaying repetitive junk. Incidentally, the process isn't perfect, and mutations leak into the surrounding DNA. It is possibly for this reason (collateral damage, you might say) that not all related fungi have an RIP system.

The other well-resolved case of directed mutation is that of so-called *somatic hypermutation*. When we get infections and our bodies mount an immune response, we have a set of cells (B cells) that have an extremely high mutation rate in one particular part of one particular gene: the variable domain of the immunoglobulin component of the B cell receptor. You can think of the gene as encoding a key to a lock, the lock being a protein belonging to the infectious agent (a virus, bacterium, etc.). We make millions of these cells, and each is mutated differently at this key gene—the mutation rate at this gene goes up about a million times the normal rate. It is indeed not the stem of the key, but the variable bit that goes into the lock that has the high mutation rate. The lock remains the same (we can't affect that bit), but we are trying out millions of different shapes to the ends of the key. When we find one that works, we expand the population of cells with the working version of the key, and our immune system can now go to work clearing out the intruder.

We don't transmit the mutations to our children, as they are in B cells, and B cells don't make sperm or eggs. Hence, we call this "somatic" hypermutation. As we don't transmit the mutations, this example too doesn't challenge the conventional view of the adaptation process.

A broadly analogous system that is, in biotechnological terms, revolutionizing biology and medicine is the CRISPR-Cas9 system. This is a bacterial system that specifically targets phage DNA (i.e., that of a bacterial virus) for destruction. The bacteria concerned have small sections of their DNA that match the DNA sequence of their past viral invaders. These sequences sit next to each other in the bacteria's DNA—they are said to be clustered, separated by flanking sequences that read the same backward as forward (they are palindromes). The sequences were thus called Clustered Regularly Interspaced Short Palindromic Repeats—or CRISPR for short. When these are made into RNA (known as CRISPR RNA), the RNA of the section that matches the viral sequence is used to guide the CRISPR-associated protein 9, Cas9 for short, to the corresponding complementary DNA of the virus. Cas9 then cuts the viral DNA, so inactivating it.

Tweaks to the system now mean that it is possible to identify a specific small section of DNA in our genome (rather than a bacterial virus) and guide the DNA-snipping enzymes to that location, using a designer guide RNA. With this we can inactivate genes that we don't want to be active. This is proving to be a very useful tool to understand what each gene does, but it also, as we will see in the last chapter, has applications to medicine: it is already being used to cure individuals of sickle cell disease.

As with RIP and somatic hypermutation, this CRISPR targeting of viral DNA works because the mutations—or cutting events in the case of CRISPR—cannot make things worse. In the case of somatic hypermutation, genes encoding for the keys start out as any old key—they are unlikely to be brilliant to start with, and are thus unlikely to have harmful mutations in them: a key either doesn't fit or does, and if it didn't fit before mutation, it can't be any worse after mutation. For other genes, e.g., those whose protein products bind DNA or process glucose, as the thing they bind to doesn't change, they evolve to be ever so good at their task. For them, mutations are more likely to disrupt. This peculiarity of RIP, CRISPR-Cas9, and somatic hypermutation may well explain why directional mutation is otherwise poorly evidenced.

The extreme idea of directed mutation would be one in which, say, all white moths sense that the trees are black and specifically mutate just the color gene, just in a way that makes them black. Were this sort of adaptive mutation the case, the population could change overnight and be hard fixed for the genetic change. Differential birth and death of black and white moths would not be part of the equation. One might wonder how exactly the moth could be so "smart" in the first place, but that is another story. The walls of the Darwinian Jericho would tumble if that was generally the case. It isn't.

To be clear, such a process is different from the adaptive plasticity of, for example, the chameleon that is adapted to be able to change color. Such color change is not manifested by a change to its DNA each time it changes color.

Others have suggested that certain parts of our DNA are more mutationally protected (i.e., have a lower rate of spontaneous mutation) than others, the protected parts corresponding to the more important parts. This is the weakest form of non-random mutation. This idea has cropped up sporadically more or less every twenty years since about 1970. There isn't much good evidence for it. We do see mutation rates higher in bits of DNA that are more superfluous (junk), but this probably has nothing to do with directing adaptive mutations as such. In the lab weed *Arabidopsis*, for example, one way the plant shuts down jumping genes is to smother them—at the DNA level—in chemicals that stop the jumping gene from jumping. A knock-on consequence of these chemical additions to the DNA is that the DNA mutates more. Not a lot more, just a bit more. In *Arabidopsis*, jumping gene DNA then has a mutation rate a bit higher than flanking intergenic DNA. This looks a bit like RIP, but the rates are far lower, and the jumping genes are not inactivated in the short term by the mutations, they are inactivated by their chemical coating. You don't have to invoke anything special to explain this, and it again looks to be a side effect.

In between the extremes of passive increases in the mutation rate owing to chemical suppression—as in *Arabidopsis*—and imaginary systems that direct mutations to just the right gene, at just the right place, at just the right time, is the idea that some adaptive mutations are created at a higher mutation rate when the mutation would be adaptive. Such a possibility still requires selection to cause the mutations to go from rare to common, so in that sense they are still within the fold of the normal model. But nonetheless, there would be a directionality in the mutation process that is not in the standard model. The HbA/HbS system has been suggested to be a rare (so far unique) example of this.

Researchers selected 6 base pairs that spanned the HbS mutation and, as a control, a 6-base-pair window in a different globin gene, one not implicated in malaria resistance. The strange claim is that in Africa the germline mutation rate in the 6-bp HbS window is higher than it is in Europe or in the comparable other gene. The rate of the key mutation (replacing glutamate with valine) is thus claimed to be especially high, but only in the environmental context where it could be adaptive (Africa).

At first sight this challenges an assumption that mutation is random, and selection is the directing process. If the claim is true, here the mutational process seems to be more in the direction of the outcome that is, in the specific context (Africa but not Europe), selectively favored.

I am not convinced. To come to these conclusions the researchers looked at many sperm, but from only 11 men. They report 9 HbA to HbS mutations, all in the 7 Africans in the sample, none in the 4 Europeans. Aside from the low numbers, there are more mutations than individuals. Only 3 people showed the mutation: 1 had it detected in 5 sperm, and 2 others had it in 2 each. Each occurrence in the sperm of the same man is most probably one mutational event. The actual number of new mutational events (as opposed to samples containing the mutation) is thus 3 in Africa and 0 in Europe. Remember too that there were also more Africans in the sample, so we expect more to come from the African sample for that reason alone. We can do some statistics on these numbers, but the 3:0 difference with 7:4 individuals is not even close to being significantly different from what you would expect by chance. You would expect this by chance alone over 11% of the time. In addition, for reasons unknown, the mutation rate across the genome is generally higher in Africa than in Europe.

If the mutation did have such a high mutation rate as claimed, we might also expect the HbS mutation to have been independently derived on several occasions. Recent work has sought to uncover the evolutionary history of the locus and to address this question. Larger-scale studies using complete genetic information now all seem to agree on the single-origin view. The age of origin is not certain, but 22,000 years ago in Africa is the current best estimate. I remain skeptical of the directional mutation claims.

Aside from somatic hypermutation, RIP in *Neurospora* and CRISPR-Cas9-directed mutations then seem to have no sound evidential basis. I would, however, be a bit surprised if there was no relationship between the rate at which a mutation happened and its role in adaptation. Do you recall Haldane's sieve? This is the idea that the adaptive mutations that we get to see are those that are dominant ones (effects visible when in old/new combination), because they have an immediate advantage when rare. As regards the mutation rate seen at adaptive sites, we might also expect that the adaptive mutations that we get to see are also those that, for whatever reasons, are not simply dominant but also happen more often.

We see some evidence for this connection between mutation rate at a given site and occurrence during adaptation in bacteria. Possibly the best example of this has been revealed by my colleague in Bath, Tiffany Taylor, who studies the bacterium *Pseudomonas*. Normally this bacterium can swim. It has what looks like a spiraling corkscrew at one end. Turning the corkscrew allows it to move about. She knocked out one of the genes necessary to make the corkscrew (technical name flagellum) and put the bacteria in the center of a dish where there wasn't much *Pseudomonas* food. However, there was more food in the near distance, not quite at the center. This experimental design should select for those bacteria that, via mutation, compensate for the knocked-out gene, make their flagellum, swim to the food, and so proliferate.

What she found was rather surprising. Repeatedly, when she did the experiment, within 48 hours a subpopulation of bacteria had evolved that make a flagellum (you can perhaps see why the evolutionary world loves bacterial evolutionary experiments—fast and cheap). What was most surprising, however, was that the bacteria did so each time with the same mutation in the same gene.

How come she saw the same mutation every time? It turns out that this isn't the only mutation that can enable the recovery of the flagellum.

But the other mutations in her species happen at a slower rate than the one she always sees. The repeatability is owing to a funny configuration of the DNA in the vicinity of the site that mutates that causes it to have a very particular high mutation rate. The rate doesn't go up just because it is needed. It just happens rather fast, and so, when populations adapt, they use the same adaptive mutation each time. Interestingly, if you change the genetic background (a slightly different strain of *Pseudomonas*), we don't see that same mutation being the one involved. This is because the mutation rates of the different mutations that could rescue motility change with genetic background.

How common this sort of bias might be, and how commonly we see higher mutation rates for adaptive genes more generally, remains to be seen.

Don't Waste Your Money, or: Why There Is Good News . . . and Bad

So far, we have largely considered the problem of the mutation rate when offspring get to inherit the mutations. This can be either through cell division in a unicell species, or via sexual reproduction in multicellular ones. This has brought us only bad news: a high mutation rate and a heavy burden of genetic disease. But what about mutations that cannot be inherited? Until recently there has been very little to say about these, as their rate is so hard to measure. What is emerging, however, is a very different story from the problem of mutations that can be inherited. And, finally, some good news for us humans. Which may also be bad news.

The major issue for such somatic mutations is that it costs a body to keep the rate low. These are costs of making repair enzymes, and of doing cell division slowly and carefully. But what about a cell in a petal of a peach tree, for example? The delicate, beautiful petal will last a few days, weeks at most, before gently floating to the ground. Why waste your energy on keeping the mutation rate down in such an ephemeral structure? These mutations, unlike germline mutations, will literally fall to earth in no time at all. Unlike germline mutations, somatic mutations have no evolutionary future.

With this thought in mind, with my co-workers Sihai Yang and Long Wang, we determined the mutation rate in peach petals compared to leaves. Leaves come from the same underlying bunch of cells in plants but persist much longer. Indeed, we found that petals have a very high mutation rate. In fact, petals seemed to have an off-the-scale mutation rate. More generally, we found that the future longevity of a cell's descendants predicts its mutation rate.

This all makes sense in economic terms. It is the same principle whereby people don't wash rental cars. If you only have the car for a few days, why look after it and spend good money on it? If you lease it for many years, however, you have more incentives to look after it.

What then would you expect to be the mutation rate in a somatic cell in a mouse versus a comparable cell in a human? The problem with somatic mutations is that not only does it cost you to prevent them, but if you don't prevent them, you get cancer. Or your cells just die. There is then an obvious trade-off. If you are a short-lived species, you care more about saving on the repair bills and instead investing the resources in making kids. A mouse will be dead in two years no matter what. Bodies are transient things. For humans, we could not hope to even make it to reproductive age if we had a mouse's somatic mutation rate. For us, then, the balance shifts the other way—spend the money (energy) to keep the somatic mutation rate down to stay alive longer, have kids, and look after them (for a bit).

Across species, then, we expect a general trend whereby somatic mutation rates can be high when life expectancy is short, and lower when life expectancy is longer. Until recently this has been next to impossible to address on technical grounds. Recent work has broken this limitation and confirmed this prediction. Somatic mutation rates go up as longevity goes down (fig. 6.5).

The researchers looked at 16 different mammals and worked out the somatic mutation rate and what you could think of as the total burden of mutations at around the time you die. The species they looked at varied 30-fold in lifespan and 40,000-fold in body mass. There was huge

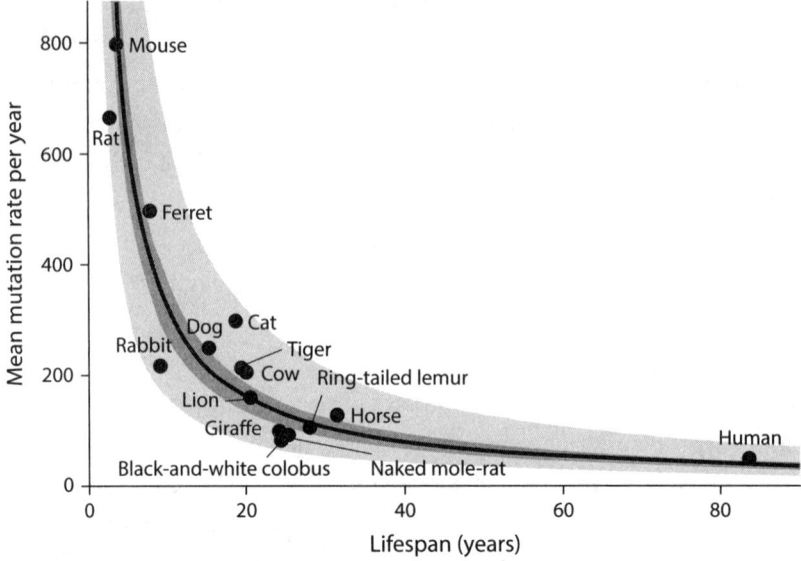

FIGURE 6.5. **Somatic mutation rates per year predicted by lifespan.** Adapted from: Cagan, A., et al. 2022. "Somatic mutation rates scale with lifespan across mammals." *Nature* 604:517–524.

variation in the somatic mutation rates (fig. 6.5). Despite this huge variation, the somatic mutational burden at the end of life was approximately the same in all species. This was because long-lived species had invested more in repair and so had lower rates. Lower rates over a long time generates the same number of mutations as higher rates over a short time. All bodies are, in a sense, rental cars, but if we have the car for longer we do have an incentive to spend money to make sure it keeps working.

Why is evolution on somatic rates then so different from the evolution of germline rates? For germline rates, we humans have the highest. For somatic rates we have one of the lowest (some long-lived plants probably have even lower rates) per year. The key difference is that germline mutations can in principle go on forever. If our bodies are rental cars, the germline is the house we bequeath to our children. We look after it for their sakes. How does this square with our high germline rate? We may have a high germline rate, but the best evidence is that per cell division it is still lower than our somatic rate. Somatic rates are under selection to be kept down, but as they have a limited future (all bodies are like petals in the end), the bigger factor is not effective population size but rather is simply the stronger selection on staying around to reproduce and look after the kids. This is not in the domain of weak selection. There may be a second-order effect whereby we have a mutation rate a little bit above what is best for us (because of a small population size), but this is minor compared with both the strong selection to balance costs and the benefits of keeping the rate low—and has yet to be investigated.

This trend for low somatic rates in humans, meaning we don't all die of cancer before we reproduce, is the one bright note when it comes to mutational evolution in humans. It also suggests that we evolved to have a somatic mutation rate fit for a species with a rather shorter life expectancy than we currently have. Increasing cancer rates in the old are an almost inevitable consequence of us not dying younger (as we used to do). Yet more bad evolutionary news. Sorry. On the slightly more positive side, the rate for a species that lives to 50 seems not greatly different from that for a species that lives to 80.

Fighting Back: Manipulating Mutation Rates for Our Good

For the most part, the evolutionary news for humans as regards mutation rates is not good. We have a high germline rate, giving us a heavy burden of genetic diseases. While our somatic rates are low, we are probably living too long for them.

This is all because, as we have seen, mutation is mostly bad for you. Folks who died from radioactivity exposure are (or, rather, were) a testament to this. Radioactivity is a potent creator of mutations. It is also why you shouldn't sit under UV lights to tan yourself. We can, however, use the same insight to improve our lot, beyond giving sound advice to avoid Fukushima, Chernobyl, and sunbeds.

Aspirin and other nonsteroidal anti-inflammatory drugs may help to prevent cancer and have been recommended as a preventive therapy. The mechanisms by which aspirin might work remain unclear and are likely to be complex. Analysis of colorectal cancer cells grown in the lab suggests aspirin reduces the rate of cell division of potentially cancerous cells while somehow promoting the killing of tumor cells. Other evidence suggests aspirin inhibits blood vessel formation, necessary for a tumor to thrive. It has a direct effect owing to its ability to reduce inflammation, as chronic inflammation is somehow coupled with cancer.

One further possible mechanism, however, is that aspirin may reduce the mutation rate, so directly reducing the chance that a normal cell might become a cancer cell. One group of researchers worked out the rate of mutation in a precancerous condition, Barrett's esophagus, and compared it between those taking aspirin and those not. Some of the patients started aspirin and were then taken off. Some started without aspirin and were then given aspirin. What was found is dramatic: the mutation rate was ten times lower when people were taking the aspirin compared to when they were not.

How it does this is not clear, but aspirin can increase levels of the mismatch repair proteins in colorectal cancer cells. Another likely route is via reduced inflammation. During inflammation we produce reactive oxygen and nitrogen species (RONS), such as nitric oxide (NO), both to combat parasites and to stimulate tissue repair. However, RONS can also damage DNA, so causing mutation. This effect is a simple chemical interaction between RONS and DNA. RONS also impede some repair enzymes. Unfortunately, DNA damage also initiates more inflammation, so an inflamed tissue can get caught in a never-ending cycle in which inflammation stimulates RONS, which makes for DNA damage, which stimulates more inflammation, etc. Something like aspirin to break that cycle might help to reduce the mutation rate along the route.

Whether one should take aspirin as a cancer preventive is, however, not crystal clear, even if it does reduce the mutation rate. The suggestion that aspirin might reduce cancer incidence led to clinical trials in which individuals were randomly allocated aspirin or a placebo without their knowing which. This allows for the fact that in the observational studies it may have been those perhaps more health-conscious who took the aspirin. Included in these trials were some on people with Lynch syndrome, a condition associated with raised cancer risk, especially of colorectal cancer. Strikingly, in one such trial those on aspirin (high dose) for at least two years cut their colon cancer risk by more than a third. On the back of such trials, in 2016 the US Preventive Services Task Force recommended daily low-dose aspirin, as it reduced the risk of getting or dying from colorectal cancer.

However, all silver linings have a cloud. A subsequent randomized controlled trial, "Aspirin in Reducing Events in the Elderly" (ASPREE), found that for people over seventy low dose aspirin *increased* chances of being diagnosed with an advanced cancer or of dying from cancer. Since the 2016 recommendation, the evidence has more generally become rather confused. Indeed, a 2022 report on all randomized controlled trials for colorectal cancer found that the evidence that aspirin was preventive, even if taken fairly early, was not strong. There was some evidence, however, from two longer-term trials that aspirin was associated with reduced mortality from this cancer. Unfortunately, aspirin use is not without risks: in 13 trials low-dose aspirin was associated with a 58% increase in bleeding in the gut and a 31% increase in a bleed on the brain. Because of this, the recommendation to take low-dose aspirin as a preventive has been curtailed. It may be a case where identification of individuals at risk—for example, if you have older brothers or sisters with certain cancers—might help. It is very much a case of talk to your doctor.

Conversely, we can apply the same principles to rid us of cellular visitors we don't want. Death by mutation is the principle behind Merck's anti-COVID drug molnupiravir. This drug is unusual because it doesn't try to block any key biochemical pathway, as most drugs do. Instead, it increases the mutation rate of the virus. The idea is that if we can ratchet up the mutation rate high enough, then the virus would be like a plane flying through a sky with so many bullets that it stands no chance. We talk of a mutational threshold, or error catastrophe, over which we can push the virus, for which obliteration is its only future. Lethal mutagenesis has also been suggested as a therapy for HIV and for some bacterial infections.

There has, however, always been a problem with lethal mutagenesis as a treatment option—aside from causing mutations in us or our unborn kids (molnupiravir is not recommended for pregnant mothers or women trying to get pregnant). What if the induced mutation rate is not above that threshold? What would then happen? Normally, you might think that you would just have a heavily damaged virus instead. A plane with lots of bullet holes may just be able to return to base. However, viruses play cat-and-mouse evolutionary games with us. Might we not have just accelerated the virus's ability to counter-evolve against us?

Life at these high mutation rates we think is odd. When thinking about evolution, we often talk in terms of the survival of the fittest. Think of the fittest as the type that is at the top of a very sharp fitness peak (in our "fitness landscape" language). Implicit within that idea is that the best may occasionally mutate, but mutation isn't so commonplace that the population cannot stay at the peak. A mutation occurs in one individual, selection eliminates it, the population is perfect again.

At higher mutation rates, this logic can no longer apply. Mutation is then so common that selection cannot sustain a population at the highest fitness peak. Mutation pushes everyone off the peak. Where then does the population go?

What is seen theoretically is that when mutation rates are high, the population will tend to become an inter-mutating cloud of types where mutations convert between forms of about the same—but low—fitness. The best plane is no longer the one that is fastest or most maneuverable; selection now favors planes that are robust to being hit by loads of mutations. And so, the sky fills up with planes with bullet holes in different places but all able to fly. This idea breaks our conventional thought of what a species is, and so the mutational cloud of types is called a quasi-species. Physicists describe an electron not as a particle but a probability density function. So too with quasi-species: not one thing but a distribution of things.

In this strange world there is no relevance to the concept of the fittest individual, as "individuals" do not hang around long enough (the unrobust speedy planes get shot down). As mutation rates go up, populations move from a mode of evolution that looks like survival of the fittest (low mutation rates) to survival of the inter-mutating cloud (high mutation rates), with a merging of the two at intermediary rates. Only when we get to very high rates do we get to cross the threshold that ensures that nothing survives. This concept of the survival of the inter-mutating cloud is sometimes called "survival of the flattest."

The concept of the survival of the flattest might seem like a strange world (and it is), but there is evidence that the theory works. One experiment considered competition between two virus-like species infecting a plant. One had fast growth, the obvious Darwinian winner. The competitor was a slow but mutationally more robust species. When growth conditions were favorable and mutation rates low, the fast species dominated, with little variation in the resulting population. That is standard survival of the fittest. However, when the experimenters turned up the mutation rate, then, as survival of the flattest predicted, the slow-growth species outcompeted the fast species. This population was indeed like a diverse, mutationally robust low-fitness cloud, not a single best speedy answer.

This sort of consideration is a concern for giving people molnupiravir. If the dose is just too low for the number of copies of SARS-CoV-2 in you, then you may be promoting the evolution of a highly diversified population by selecting for mutational robustness. It also seems to apply to the way some viruses evolve. HIV has the highest mutation rate of anything known (if you are interested in numbers, it is about 0.004 per base pair per cell). It probably exists more like a cloud of inter-mutating types.

With that caveat aside, with aspirin and drugs like molnupiravir we can use human ingenuity to fight back against our rotten evolutionary lot. Nonetheless, the burden of genetic diseases remains high. And this might be the tip of the iceberg. To be classified as a genetic disease, the affected baby needs to at least make it to term, to come into the world. Embryos that die before birth are not included in the 1-in-11 statistics. It is at this stage, however, that the great majority of potential humans die. In the next chapter we shall see why being a mammal only makes our rotten evolutionary hand worse still.

7

Problems with Placentas, Pregnancy, and Perfection

If you thought that between 5% and 10% having a rare genetic disease is rather a high degree of suffering for a species that many see as some sort of pinnacle of evolution, just wait until you consider what happens before birth. Human reproduction is breathtakingly wasteful.

Perhaps the most familiar problem with pregnancy is miscarriage. Of recognized pregnancies (that is, typically after week 6), one-fifth will not make it past about week 20 of the pregnancy. Most such miscarriages happen before weeks 12–13. In the UK we estimate that for every 2,000 babies born each day, there were 500 miscarriages, 150 babies born preterm, and 7 stillbirths. Indeed, so common is miscarriage that, in many medical systems, the doctors will not be interested in even thinking about trying to investigate further until you have had three or more miscarriages.

The most striking statistic, however, concerns the number of embryos that never make it to be recognized as a pregnancy. Analysis of the youngest embryos (just a little after fertilization) is now genetically possible. A remarkable 50% of these are genetically doomed to die, most without the mother even knowing that conception had ever occurred. The numbers depend quite a bit on the mother's age, but the usually quoted range is somewhere between 40% and 60%. Others claim more like 30–50%, but we are splitting hairs. These numbers are remarkably high. Current estimates suggest that for every baby born another two potential babies died.

It is not uncommon for parents who have suffered a miscarriage to be told that it was for the best. Often, they are reassured that it is "nature's way" of separating the embryos that had no prospect from the potentially fit and healthy babies. You might be told that, if the fetus has no prospects of surviving, it was better to miscarry earlier rather than later. This contains some truth: our mode of reproduction is such that if an embryo dies early, the mother is saved from further investing in that embryo. However, it also raises another question: Why on earth are so many embryos genetically so poor in the first place? Surely, after a few hundred million years of animal evolution, natural selection would have perfected the art of reproduction such that almost no embryos come into being that are not capable of making it.

Increasing evidence now suggests that these two aspects are coupled: we—and other mammals—have remarkably wasteful reproduction because, owing to placentation, we can cut short the resources going to a potential baby. More generally, other than our low population size, if there is a second big reason that we aren't at some sort of pinnacle and suffer from abundant imperfections, it is that we are also a mammal with an early dependence on our mother's womb and breasts for food. We have problems with placentas and pregnancy.

We have already seen a few issues raised by having a placenta. We saw, for example, how mothers and their babies may "disagree" about how much sugar the baby should get. The baby appears to manipulate the system to try to get the mother to give more by secreting human placental lactogen across the placenta into mum's bloodstream. This stops the mother from taking glucose out of circulation (more for the baby). This can lead to gestational diabetes. The mother counters this by increasing the number of beta cells—the insulin factories—in her pancreas to produce more insulin.

We also have seen possible tussles over the mother's blood pressure: higher pressure means more resources for the baby. This can lead to gestational hypertension (high blood pressure in the mother). Some think pre-eclampsia, the potentially fatal, human-specific condition of pregnancy that affects 5% of pregnancies, may be related to such maternal-fetal conflicts. Other evidence suggests instead that it is owing to the mother's immune system misfiring against the placenta. It is, after all, unusual to have a foreign entity in you and not mount an immune response against it. Either way, the placenta, and the intimate mother–fetus interaction it creates, is at the heart of what remains a great killer. In low- to middle-income countries, where 99% of pre-eclampsia-related deaths occur, about 16% of all maternal death is owing to pre-eclampsia (and possibly related high blood pressure conditions). That is about 76,000 mothers and 500,000 babies worldwide who lose their lives each year owing to the blood pressure problems created by the maternal–fetal interface.

More generally, the process of giving birth to a live baby is a dangerous process for mother and baby alike. The greatest killers of mothers are post-delivery infections and post-delivery bleeding. Despite a global reduction in maternal mortality between 2000 and 2020 of about one-third, the WHO estimates that in 2020 a maternal death occurred globally every two minutes. For every maternal mortality, it is estimated that there are 20–30 mothers seriously harmed by pregnancy and childbirth. In the absence of modern healthcare, the lifetime risk of maternal death in childbirth is around 5%–10%, with a rate per live birth at around 1%, although it is reported to be as high as 3.5% per live birth for the Agta peoples of the Philippines. Analysis of 128 pre-Columbian female mummies from Chile dating to about 1300 BC found that a remarkable 14% of the dead women showed childbirth complications. Pregnancy and childbirth are, and likely always have been, dangerous.

Amongst mammals, humans, it is thought, have particular difficulties not just with pre-eclampsia, which is human-specific, but also with our rather large heads compared to the size of the birth canal. In other mammals, death during childbirth is usually reported to be uncommon, birth progressing relatively easily for most nonhuman primates, for example. Nonetheless, some data suggest that other primates may have related problems. In captive pigtailed macaques, 14% of births were reported as being difficult, and in captive squirrel monkeys that number may go as high as 50%. In these, most complications seem to stem from the baby being in the wrong position in the uterus (breech births).

Our Development Enables Reproductive Compensation

While the birth process is itself problematic for humans, all mammals are likely to have placenta-related issues. The problems we have owing to the placenta are not solely related to the fact that it provides a means for mother and baby to squabble and interact. Our mode of looking after baby also means that if an embryo dies in the womb, the mother can save the food she would have given to the embryo and then redistribute it. This saving indeed applies through to the end of breastfeeding.

In this regard mammals are unusual. Let's compare the mammalian way of looking after the embryo and baby to, for example, what happens with birds. In the latter the mother gives resources to the egg—the white of the egg and especially the yolk (very nutrient-rich). Not only can the embryo and the growing chick not demand more, but if the chick dies in the egg, these resources are lost. For a mammal, if the embryo dies, the mother just stops giving up sugar and other resources. Birds will also sit on dead eggs, wasting time and energy. Human mothers, for the most part, do not. Birds cannot have a miscarriage, just a dead egg that they will still incubate. Even kiwi birds that only lay one egg will sit on a dead egg.

In many regards the mammalian way is the better way: we don't waste time and resources on dead embryos. If an embryo is to die, much better that the mother does not invest everything in the embryo in advance (as birds do), and instead redistributes the saved resources to the viable siblings.

This ability to make up a bit for the early loss of an embryo is known, perhaps rather obviously, as reproductive compensation. We can get a rough idea of how big these effects are by looking at data on human twin and singleton (just the one baby) births. At birth the latter, on average, weigh about 40% more than each member of a pair of twins. A twin pregnancy in which one embryo dies early thus realizes about 70% of the investment that would have been. Without redirection of resources there would have been half the total. From data on the relationship between birth weight and infant survival (in the absence of advanced healthcare), this change in birth weight increases the singleton baby's chances of survival about 10%.

Reproductive compensation can do odd things to selection. In mice, for example, there is a strange bit of genetics whereby a baby mouse that has two copies of a given section of a chromosome (called the t-complex) will grow up to be sterile. Because of reproductive compensation, selection favors other mutations in the same embryo that kill these embryos as early as possible in the mother. Yes, you heard that right: natural selection can favor mutations that kill the embryos they are in—and do so as soon as possible. If you think that is weird, prepare to have your mind blown.

Reproductive Compensation Makes Embryonic Lethal Mutations More Common

Perhaps the more ordinary consequence of reproductive compensation is that mutations that kill you early are expected to be at higher frequencies in mammals—and thus cause more death—than we would see in other species. Consider, for example, a mutation in a gene that causes the embryo to die if that embryo has both versions of this gene with the mutation. In the official language, the mutation would be an embryonic lethal recessive. It is the nature of such mutations that no mother or father can have two bad copies (individuals with two lethal copies die soon after conception). But they can each have a good and a bad copy. If such parents in turn have babies together, a quarter of fertilized eggs will have two copies of the bad version just because of the rules of inheritance. These babies die, possibly very early on in the womb. In all other species, that would have been the end of it, and the mutation would be in the population at the simple mutation-selection equilibrium. Not so in mammals. The death of the embryos with two bad versions of the gene increases the food going to the sibling survivors. But, given the genetics of the parents, the survivors are more likely than the average embryo to have the lethal mutation (this will be in a good/bad combination). If both parents have one good/one bad, then 2/3 of the survivors have the lethal mutation and get the benefit of the death of the embryo. In no other mating combination do embryos die. Consequently, in mammals, embryonic lethal recessive mutations are expected to be more common, as they disproportionately benefit from reproductive compensation. They will always be in the right place at the right time to receive the released resources.

Ian Hastings of the University of Liverpool has estimated what sort of effect this might have on the prevalence of human lethal mutations. He finds that, because of reproductive compensation, embryonic lethal mutations should be between about 22% and 33% more common than they otherwise would be at the expected mutation–selection equilibrium. Reproductive compensation may even be especially strong in humans, as, if a single embryo dies, the mother doesn't redistribute resources to the other members of the same brood, as mice and dogs do (which is a bit inefficient), but instead will often just rapidly reproduce again.

Whether this effect can account for the commonness of rare diseases, however, is not so clear, as, by definition, a rare disease is a medical condition in which babies survive to birth and beyond. However, if some also die early (in the womb or before weaning), then they would be associated with reproductive compensation and would exist at frequencies higher than expected. Intriguingly, a survey of lethal recessive genetic diseases by Molly Przeworski and her colleagues of Columbia University, New York, reports that recessive lethal mutations are usually more common than would be expected in the absence of reproductive compensation. However, we probably also know more about those diseases that are more common, so her disease gene sample may have an ascertainment bias toward those that are strangely common.

Reproductive Compensation and
the Wrong Number of Chromosomes

While the relevance to rare diseases isn't so clear, the compensation effect can go some way to explaining why most human pregnancies fail without the mother knowing she is even pregnant. If Hastings is right, however, the effect of recessive embryonic lethals is not of a magnitude large enough to explain the remarkably high rates of early embryonic death. This makes sense, as we also know that the great majority of these early embryonic deaths are not owing to recessive lethals. They are owing to spontaneous increases or decreases in the number of chromosomes that an embryo inherits. A gain of one chromosome is called a trisomy, a loss of one is a monosomy, and collectively they are called *aneuploidies*. They are, by quite some margin, the greatest killer of potential humans.

Nearly all (99.6%) of human embryos that are born have the canonical 46 chromosomes. That is, 23 from dad and 23 from mum. Of the non-sex chromosomes (i.e., the 22 chromosomes that are not X or Y chromosomes), rare exceptions are babies born with three copies of chromosome 21, or of chromosome 18, or of chromosome 13. The first of these is the best known, this being the most common cause of Down syndrome. The other two cause Edwards syndrome (chromosome 18) or Patau syndrome (chromosome 13). While some instances of these three trisomies can make it to term, most do not. About 80% of trisomy 21 cases die spontaneously in the womb. Most instances of the other two also die in the womb, and if not, then shortly after birth. Only about 10% of Edwards and Patau syndrome babies make it to their first birthday. With advanced healthcare, of the 20% of trisomy 21 cases that are born (i.e., Down syndrome babies), most can survive, but still, about 10%–20% die soon after birth. Without advanced healthcare few survive past their fifth birthday.

However, these syndromic cases (i.e., where the embryo/fetus makes it to be born) are the tip of the aneuploidy iceberg. With advances in genetics and with access to eggs and early embryos, for example from patients who have their eggs harvested in preparation for chemotherapy, we now have a window into what is happening at the very earliest stages of development. In these very early embryos or in the eggs, we find a gain or loss of other chromosomes (those never associated with live births) very regularly. It is by looking at the early embryos that have the wrong number of chromosomes that we know that about 50% of our earliest embryos die. In the eggs, perhaps more than 70% have the wrong number of chromosomes. In mice, no pups are born with chromosome number differences, but plenty of early embryos have them.

Most embryos with the wrong number of chromosomes die without implanting in the womb. These are never recognized as pregnancies. Of the 15%–20% of recognized pregnancies that result in miscarriage, about 35% have the wrong number of chromosomes. The very high number of embryos that die because they have one too many or one too few chromosomes is a real enigma. We have for a while known how, in mechanistic terms, the gain or loss of a chromosome happens. Only recently have we started to understand why, evolutionarily speaking, it might be so very common.

As to the how, we need to start by understanding how males and females make their sex cells, their gametes. These are sperm in males and eggs (or ova, to give them their more proper name) in females. We briefly considered this previously, but I would like to bring your attention to certain oddities. In principle, making sperm and making eggs both involve the same process, one in which a cell starts with two copies of each chromosome and ends up with a cell (egg) or four cells (sperm), each with one copy of each chromosome. The process is termed *meiosis*.

Let's start by focusing on just one pair of our 23 pairs, perhaps the version of chromosome 21 you inherited from mum and the chromosome 21 you received from your dad. We can think of them as a pair of **I**-shaped structures: **I** and **I**. Given that the aim of meiosis is to reduce the number of chromosomes, step 1 is odd. Each of the chromosome pairs is replicated (doubled) to make, in sum, four copies of the chromosome. They aren't, however, four independent copies. The two maternal ones are joined together at the hip, at a special pinch point called the *centromere*; likewise, the two paternal ones: each I-shaped structure has duplicated to now look like an **X**-shaped structure (not to be confused with X chromosomes, although in step 1 of meiosis they also make an **X**-shaped structure). The centromere is the central pinch point. Now what happens is that these two **X**-like structures pair up:

XX

Cellular ropes are attached to the two pinch points, the centromeres:

----**XX**----

And the maternal and paternal **X** structures are pulled away from each other:

----**X**   **X**----

Note to reader: Remember the centromere-rope attachment—it will be important. By the way, it isn't actually rope (strange to say), but so-called microtubules made of a protein, tubulin. But they may as well be thought of as rope.

At this point the stories for male and female meiosis are different. In male meiosis, we now have two cells, each with one **X** structure for each chromosome:

(**X**) (**X**)

In female meiosis, there is one "big egg" (technically, the secondary oocyte) that receives one **X** and a smaller cell, called the polar body, that receives the other **X**. The polar body is a genetic dead end (R.I.P). Only the chromosomes in the egg will have a future in the next generation. They will become the egg that will be fertilized by the sperm. Further note to reader: remember that fact—it will also be important. Incidentally, with eggs we can look in the polar body of the eggs to calculate the numbers of each chromosome and so infer the number in the eggs while doing no harm.

That is the end of meiosis step 1, otherwise known, rather logically, as meiosis 1. Meiosis step 2 now takes the remaining **X** structure and splits it, separating what was the prior pinch point/fusion point, the centromere. In males, one cell again becomes two cells, again with the ropes attaching to the centromere, pulling apart the ----**X**----- so that it becomes -----**I**   **I**-----, and each new cell has a single copy of this chromosome. Since two cells started this second step in males, the original step 1 cell with two copies of each chromosome, maternal and paternal, has now given rise to four cells, each with one copy of each chromosome. These are very compact cells, with the DNA crammed into the cell's head and a wiggly tail: sperm.

In females, recall there is only one cell after the first step: the big egg. This undergoes the same process and again throws one copy of each chromosome into a second polar body. This way, one original cell with two copies of each chromosome ends up as one cell, the ovum, with one copy of each chromosome. When egg and sperm meet, in principle you should now have a fertilized egg with two copies of each chromosome, one each from mum and dad.

In many instances, in the first division there is a second process that happens. When the two **X** structures meet up (**XX**), they swap over segments of DNA on the two touching branches of the Xs. This is known as *crossing over*. It is like an extra bit of mixing up of the DNA, aside from the independent segregation of the chromosomes.

Where, then, is the error made that leads to too few or too many chromosomes being inherited by the embryo? Here lies an oddity: the error is usually in maternal meiosis 1. Only about 1%–2% of the errors in humans are from male meiosis. Errors in female meiosis 2 do happen, but they are less common than in meiosis 1 and are often owing to earlier errors in meiosis 1. An analysis of chromosome problems associated with our chromosome 16, for example, found that they all were owing to some mistake in meiosis 1.

This preponderance of maternal meiosis 1 errors is one of the first big clues, it is now thought, to understand why this error might be happening.

A second clue has recently been noticed. If we look across the vertebrates (fish, amphibians, birds, reptiles, mammals), we can ask whether they all have this same issue with large numbers of chromosome loss and gain events in early embryos. One large study looked at over 2,000 embryos of zebrafish (one of the laboratory model organisms) and found *no* chromosome gain or loss events resulting from errors in maternal meiosis 1. A similar study in the laboratory toad (called *Xenopus*) also found none. In birds they happen, but are rare. In chickens and zebra finches, trisomy rates are about 0.04% per chromosome per early embryo. In most vertebrates this problem of chromosome gain and loss doesn't seem to be a problem.

By contrast, in mammals it is very common. Mammals seem to have loss or gain with a rate of ~1% per chromosome per early embryo; this is seen in cows, pigs, mice, as well as in humans. That is 25 times higher than in birds and vastly higher than in fish and amphibians.

The 1% number is a bit rough, as the chance of loss or gain goes up considerably with mother's age. This is known in both mice and humans. Older women have a much higher chance of having a Down syndrome child and have a much harder time getting pregnant. This aside, it looks like the more chromosomes you have the greater the rate of loss/gain. Cows, for example, have 30 pairs of chromosomes (so, 30 in one chromosome set, compared with our 23), and loss/gain of any is about twice the rate of that seen in pigs, which have 18 chromosomes per set. We are something of an outlier, in that for us, with 23 chromosomes, we might expect a rate of 23% of embryos if there is a 1% chance per embryo. Our rates, however, are also affected by the fact that mothers now tend to reproduce later, and a delay to mid- to late thirties for having kids greatly increases the numbers—to the 30%–60% rate, rather than 23%.

Why would the loss/gain of chromosomes be so much more common in mammals? We think reproductive compensation has something to do with it. One possibility is that this is like the effect Ian Hastings looked at, and that reproductive compensation simply allows what would have been a harmful event to be less harmful. Consequently, it can simply be more common, as selection to reduce the rate is weaker. However, recall that this effect is thought to be relatively modest. It doesn't seem to be able to explain why fish and frogs have none, but mammals have lots of gain/loss events.

A further suggestion has then come about. To get your head around this suggestion you need to think about that first division in female meiosis, and what is happening at the centromeric parts of the two X-shaped chromosomes. In this first division one centromere will go one way—to the egg—and the other will go the other way—to the polar body. But the polar body is a dead end, genetically speaking. What if a centromere "knows" that it is going to the polar body? What would you do?

Let's think about the problem differently. To reconsider maternal meiosis 1, imagine that there are two people, you and someone else. My job (I am the master controller) is to find a way to randomly select just one of you to live, the one that can go to the egg. You are one chromosome in the X shape; the competitor in this game is the other partner chromosome, also in the X shape—perhaps the two copies of chromosome 21.

I don't care which one of you is to be selected, but it must be just one of you—not none and not two. Here is my solution. I stand you both near a cliff edge. I blindfold both of you and throw out two ropes. One is a rope of life, one is a rope of death. You catch one, the other person catches the other. I now pull on the two ropes—the rope of life drags the person with it to safety (the egg), the rope of death takes the other person over the cliff (off to the polar body). I have successfully found a way to randomly get just one of you. That is what maternal meiosis 1 is supposed to achieve and, metaphorically, how it works. My job is done.

Yours, however, is not. Are you happy with this setup? Perhaps you might be tempted to peek under the blindfold, work out which rope is which, and, if you have the one that would drag you off the cliff, to hurriedly try and rip the rope of life from the other person and make them take the rope of death. Good strategy—you win. We have a term for this strategy: it is called *centromeric drive*, and we see it commonly. It is well described in mice, for example. So long as there is still just one of you at the end of your shenanigans, I might be OK with that.

But what about this: What if you take the rope as given and, if you detect that you are being dragged toward the cliff (the rope of death), you now make every attempt to drop that rope and try to grab the other rope, the rope of life, but you don't quite succeed? The calculus is interesting here. If you knew you were facing your certain death, then actually you cannot now do any worse. You were going to die—what is the worst that can happen to you? If you don't succeed in your preferred solution (you take the rope of life, the partner gets the rope of death), there might be two of you left over in the egg, both grabbing hold of the rope of life. Or perhaps you just stay on the cliff edge and don't get taken over the edge. In these cases, I—as the master controller of organized chromosomal division—will not be best pleased. Remember, I want just one of you to survive. However, and here is the oddity, if there is reproductive compensation, you now win by killing the egg!

It is odd to think that embryo death by chromosomal gain or loss induced by a centromere could be a winning strategy. For the chromosomal gain case, you are saying, "If I am off to my death in the polar body, I'm staying in the egg to kill it." The chromosome loss case is you saying, "If I am off to the polar body, you, my partner, are coming with me, and the egg will die." You win both ways, because the egg's death forces resources to be reallocated, and you gain from the reallocation. In effect, having messed up the first round of the game, you have forced me, as master controller, to play the same game again. To be precise, you don't win—you are, after all, in a dead egg. But your identical twin (centromere) playing the next round of the game does win, so between the two of you, you win.

That somewhat bizarre insight needs some unpacking. While it might seem peculiar that this could ever be favored, a simple numerical example makes the point. Imagine a mother will raise only two children. Through early meiosis 1 our Machiavellian centromere is the "lucky" chromosome half the time, so on average goes to one of the two offspring. It was given the rope of life. Thus, if it does nothing its rate of transmission is 0.5. Imagine, however, that during the half of the time that it would not be transmitted (it was given the rope of death), it kills the surviving embryo by dropping its rope, perhaps, making the embryo have one too many chromosomes. That embryo dies. In humans, the mother, having saved resources and time, now rapidly reproduces again. This time, the same centromere (i.e., the identical relative of the rope-dropping centromere) has a further 50% chance of being transmitted, giving a net average occupancy of 1.5 of the two children. If it had simply accepted its fate, it would on average be in just one of the two kids. It has become more common just by being cunning.

It might also be the case that if it is unsuccessful in this with the third embryo, then it can kill again and again until the two surviving progeny are reared. It is then transmitted to both of the surviving progeny. Once you know you have the rope of death, with reproductive compensation you can only win by dropping the rope, especially if you kill the embryo.

There is a similar logic that applies to species with multiple babies in a brood (e.g., mice, dogs, etc.). In a mouse the mother may have, say, 10 pups in her womb. The Machiavellian centromere will on average be in 5 of these with a normal chromosome complement. The other half are, however, killed by the same centromere. The mother now hands over the resources that would have gone to the embryo to the survivors. As these survivors have the Machiavellian centromere, it wins.

When we say it wins, what we actually mean is that when this variant arrives in the population, it will increase in frequency—just as a gene variant making a moth black increases in frequency when rare if the trees are black. In the current case, however, our centromere isn't increasing in frequency because it is doing anything good for us, the bearers of the centromere—in fact, quite the opposite. But the same principle still applies. We are interested in the fate of new mutations when rare, the first appearance of Machiavellian centromeres being a form of mutation. If they can increase in frequency when rare—for whatever reason—we expect to see more of them.

You can compare this to what would happen in, say, a fish. Fish usually make eggs that are released by the female into the water; the male sprays sperm all over, and the fertilized eggs then drift off. What would happen here to a chromosome that dropped the rope of death? It would not be any worse off. But it would also not be any better off. The death of that embryo doesn't lead to the mother reproducing again any faster. As the young are just released into the water, it does not free up resources that the other embryos would then take up. There is no placental transfer of sugar, or breastfeeding after they hatch. Interestingly, in birds there may be a weak advantage. Although the mother will not reproduce any sooner, and all the yolk was already given up, with fewer mouths to feed (literally) there may be some small savings to be made. It is then suggestive that fish have the lowest chromosomal error rate, birds a small but detectable rate of gain/loss, and mammals the highest rate.

There is one mammalian exception: inbred mice don't have high rates of loss/gain. But in inbred mice, the two partners in our cliff edge game are identical twins. If you are one of a pair of identical twins, then you should take the rope you are given, because no matter who grabs the rope of life, it is either you or a centromere that is genetically identical. Put differently, a mutant centromere in an inbred species that goes around killing embryos cannot increase in frequency when rare, because with each embryonic death it is killing itself twice over, and the redirection of resources cannot make up for that. The same logic explains why the process is particular to maternal meiosis 1. Only in meiosis 1 are there two different people's centromeres (or different people in the rope analogy). In meiosis 2, the division at the centromere is splitting identical centromeric regions (either both maternal or both paternal in origin). This is because the crossing-over process is always suppressed where the ropes attach.

This is all fine in principle, but is it not just so much fantasy? Chromosomes surely cannot do anything that scheming, that apparently thoughtful or Machiavellian. But they can. They aren't, of course, thinking, but that doesn't mean chromosomes cannot behave in a conditional manner. Remarkable work by Michael Lampson of the University of Pennsylvania has indeed shown that they do. He has found in mice that if a centromere is being dragged to the polar body, this centromere can "flip," that is, detach itself from the rope of death and try to become oriented to grab the rope of life. Flipping events are strongly biased to detach larger centromeres from the rope of death and re-orient to grab the rope of life.

How, you might well wonder, could they ever "know" which rope they had (the equivalent to peeking under the blindfold)? Lampson and his colleagues have shown that across the egg is a gradient in a chemical that attaches to the ropes (the technical term is tyrosination of tubulin, tubulin being the rope's twine), which then modifies the ropes to differing degrees. Centromeric attachment is modulated by the degree of this chemical covering on the ropes—you didn't have to lift the blindfold, you just needed to feel how much of the chemical is on the rope. Or that the rope simply becomes more slippery. The first time I met this phenomenon my jaw dropped. These are just chemicals interacting, but behaving in a Machiavellian, apparently scheming way.

Notice too that, for many of the means by which a player tries to "game" the system, the player can be better off on average, but I—as game controller—often lose. For example, every time the flip doesn't work out properly and we end up with one extra chromosome, then the offspring is worse off—but the player is still better off flipping, as it has nothing to lose and will gain by compensation. So, I, as master controller, should try to stop the flipping, stop you from trying to cheat my well-devised system, and definitely stop killing embryos—the ones that were not going to transmit you onward to pastures (embryos) new. This we also see. Lampson has, for example, also shown that there are variants of a protein that binds at the point where the centromere and rope attach that suppresses the centromere from dropping the rope of death. In mice, reduced dosage of this protein (called Bub1) is associated with increased rates of chromosome gain/loss events. Bub1 levels also decline with maternal age, as rates of loss and gain go up. Human aneuploidy is also known to be owing to issues centromeres have grabbing the ropes. It all seems to add up. The idea is new and needs to be more fully tested. It isn't clear, for example, whether the chromosomal behavior observed by Michael Lampson in mice is seen in other species. Fruit flies also seem to have oddly high rates of aneuploidy, not obviously predicted by this model. But it is the best explanation for why we have so many embryos doomed from the get-go.

A further corollary of this sort of idea is that our genomes are forever playing cat-and-mouse games with our own centromeres. The rest of our genes benefit from suppressing this harmful induction of embryonic death. They will be under selection to suppress the "cheating" of the game somehow. In turn, the centromeres are under selection to cheat again. Consistent with this we see that the proteins that interact with centromeres are fast-evolving, with $Ka/Ks > 1$. In turn, the centromeres evolve super-fast too. If you think about it, that is odd—this is a key structure to guarantee that cell division and sex cell production is a flawless process. Why isn't there just a best solution that is then highly conserved? This sort of conflict is our best explanation for the rapid evolution. In fact, when people first did scans across the human genome to look for places where variants showed all the footprints of going from rare to common rapidly, they discovered lots of centromeric signals. As centromeres don't have any protein-coding genes, this sort of Machiavellian centromere is our best explanation. The way we do maternal meiosis 1 is asking for trouble, as we have reproductive compensation.

What is odd in this case of cat-and-mouse game is that this isn't us playing cat and mouse with viruses and parasites, not males and females, not babies and mothers, but instead our own genes and our own centromeres. Because of this, this is known as an *intra-genomic conflict*, or genomic or genetic conflict for short. What is best for the centromere—try to grab the rope of life if given the rope of death—is not what is best for all other genes in the genome. They, strange to say, prefer not to be in a dead embryo and would favor a neat 50:50 segregation. Incidentally, we don't call them Machiavellian, rather we call them "selfish genetic elements." Some also like the term "selfish gene" or "ultra-selfish gene." Or "genetic renegades," "genomic parasites," or "self-promoting elements." The language is diverse, but the underlying concept is the same. Some mutations increase in frequency not because they are good for us but because they game the system of transmission.

The Case of the Gene That Knows Its Parents

Placentation and the provision of milk after birth are thought to lead to another conflict in us. We saw earlier how our placental upbringing causes conflicts between mother and fetus over sugar levels that lead to gestational diabetes, that it can create the conditions for raised maternal blood pressure. This same flexibility of food delivery is thought to explain a closely aligned conflict that underpins certain strange genetic diseases.

Perhaps you have seen those images of ducklings following a person with a bucket. Where the person with said bucket goes, the ducklings follow. This is known as behavioral imprinting. When a chick hatches it usually gets to see its mother. An imprint in the brain is made of the mother, and the chick now knows to follow the mother. If we replace the mother with a bucket, the chick instead imprints on the bucket and follows that.

We have some genes that are a bit like this. I need to explain. You have about 20,000 protein-coding genes and a good number of non-coding genes. When either sort of genes are expressed, usually both copies—the one from mum and the one from dad—are expressed (i.e., transcribed and potentially translated). However, for some genes this rule doesn't apply.

In some cases, within a cell, we randomly pick one of the copies and only express that one. We do this often when cells need to specialize.

For example, up your nose a cell will typically express only one of two copies of certain olfactory receptor genes. B cells of our immune system typically only express one of the two copies of immune system proteins. Each such cell could have been a more generalist cell producing both versions, but they produce just the one. However, for about 100–200 genes, only one of two gene copies is expressed but it is not a random choice—it is dependent on which parent you inherited the gene from. For example, we all have two copies of a gene called *insulin-like growth factor 2 (IGF2)*. However, as embryos, in certain tissues you express only one of them—but always, in all of us, it is the one we inherited from dad. The one we inherited from mum is inactivated. There is a neighboring gene on the same chromosome, one that doesn't code for a protein, called *H19*. *H19* is only transcribed from the chromosome you inherited from mum. The version you inherited from dad is inactivated. It is because these genes seem to know who their parent is that they are said to be *genomically imprinted*.

Imprinting (for short) is one of those weird head-scratching evolutionary genetic problems. If you, like me, are somewhat anxious about flying, you will understand the problem. Having and using both copies of our genes is a bit like flying in a plane with two engines. Should one engine fail, you can still make it safely home. Using only one of two is then weird. You take off with two engines but only switch on one, without the option to switch on the other. You are flying on one engine. What if that engine fails? In genetic terms, if you have a recessive lethal mutation in one of the two genes, expressing both means you don't suffer the disease. You survive. If you express only one, you have a 50% chance of dying. Similarly, if you express only one and that one gets a mutation (after takeoff, so to speak), then again you could die. It doesn't make sense.

The leading explanation for this enigma was suggested by Tom Moore, now at University College Cork, Ireland, and David Haig, of Harvard University. Their logic runs something like this. Imagine you are a gene in an embryo, and down the line, your father will not be the father of all the children of this same mother (on the average, you don't need it for every case—I do not wish to impugn any readers' oath of monogamy). How much resources should you be demanding from your mother? What Moore and Haig noticed was that the answer depends on whether you are the gene inherited from mum or dad. The exact copy of the one inherited from mum has a 50:50 chance of also being in any given future sister or brother, because it has come from mum. But if there is multiple paternity (as switching of fathers is known), then the same is not true for the one coming from dad. If the father down the line is different, the exact copy will not be in the next half-sib. There would be funny exceptions if the next father was the current father's identical twin, but let's not go there.

Just as we saw earlier, Bob Trivers pointed out that this sort of difference can explain why mum and baby disagree on resource allocation (baby wants more, mother wants to hold some back). What Moore and Haig argued was that there can also be a difference between the maternally derived and paternally derived genes in the baby. The ones expressed from the paternal chromosomes should be the ones saying, "Give me more food, the next baby isn't my relative," while the ones expressed from the maternally derived side should be saying, "No, not that much, save some for more future brothers and sisters."

In mice there is indeed an imprinted gene called *insulin-like growth factor 2 receptor*. This is expressed off the maternal chromosome set and stops *Igf2* from taking the resources. The Igf2r protein binds to Igf2 protein and takes it off to a cellular recycling bin before Igf2 protein can do its growth factor business (there was a clue in the name of the gene). The growth effects can be because of expression in the embryo or in the placenta, an expected hot spot for an embryo to obtain more from mum. Many imprinted genes are indeed expressed in the placenta, and many do affect growth of embryos, in some part because of their activity in the placenta.

It isn't necessarily more food in the womb that is the only point of potential conflict between the maternally derived and paternally derived genes. The same logic applies after birth to how much milk to take. There could be genetic "disagreements" after birth about, for example, how aggressive to be at suckling. At least six mouse imprinted genes affect suckling behavior or milk acquisition. For example, in mice *Zdbf2* is expressed only from the paternally inherited gene. The gene's function seems to be to motivate the newborn mice to actively demand food (milk) from the mother right after birth. It is expressed in the young mouse brain, and failure to express it leaves the mice with milk deprivation. The pups lose weight and many die soon after birth. This fits with the idea that paternally expressed genes should be stimulating growth and obtaining resources from mum.

More generally, there could be conflicts about how aggressive or pleasant to be to sibs. Many (if not most) imprinted genes are expressed in the brain and affect social behavior. There could also be conflicts concerning whether to help keep one's brood mates warm. Mouse pups will congregate together to share warmth, for example. Haig has suggested that there will be conflicts between maternal and paternal genes as to how much to take from this communal warmth and how much to contribute. Some mouse imprinted genes appear to regulate fat metabolism and contribute to this.

This is by no means the only explanation, but it does potentially also explain why genomic imprinting is seen in mammals and flowering plants and not more generally (there may be other examples, in bees, for example). Flowering plants are a bit like mammals in that the seed can "ask" for more resources from the mother plant in early development. A chick in an egg cannot do this. Birds don't seem to have genomic imprinting. This being said, they do compete for food after hatching, and the model would predict imprinting for genes associated with such behavior. This has yet to be observed.

While this data all looks as though it makes sense, the "cleanest" set of predictions of the conflict model concerns the effects on growth of having more than or less than the usual amount of product of any given imprinted gene. The ones expressed from dad's chromosomes should say "Give me more resources," so having more of that gene expressed should give big babies and underdose should give small babies. Conversely, expression of mum's genes in the baby should say, "Hold back, let's save some for future kids." Overexpression of these should lead to smaller babies, underexpression should lead to bigger babies.

We can test these predictions using so-called gene knockouts in mice. Here what we do is take the gene away (literally take it out of the DNA or majorly break it) and see how the embryo changes when this deletion is inherited from mum and not from dad, and vice versa.

There are at least 13 imprinted gene knockouts with embryo growth effects. In 11 of these the direction of the growth effect goes with the predictions: deletion of the paternally expressed ones leads to growth retardation of the embryo, while deletion of maternal ones leads to a bigger pup. *Igf2* in mice is like this, for example. An embryo missing the gene when the deletion came from dad is small. An embryo missing this gene when it is inherited from mum is normal-sized. The gene wasn't going to be expressed anyway. *Igf2r*'s knockout is the opposite. If a pup inherits the deletion from mum, then it ends up big: remember the idea was that this should be part of a response saying, no, don't give me that much food. When inherited from dad, there are no major effects as it wasn't going to be expressed anyway. What is especially elegant is that if you make a mouse embryo with both the maternal deletion of *Igf2r* and the paternal deletion of *Igf2*, then the two effects cancel out and the pups are a normal size.

There are, however, cases in which the predictions don't so obviously get it right. Knockout of the gene *Phlda2* is associated with growth re-striction when the deletion comes from the mum. Similarly, maternal transmission of *Ascl2* knockout results in placental failure where a large, healthy placenta would be the expectation.

In humans we also see some evidence that looks to be consistent, but also some cases that don't so obviously fit. We don't make human knock-outs (that would be unethical). We must wait for spontaneous muta-tions or reports of a strange bit of genetics, so-called *uniparental disomies* (UPD). These are rare cases where individuals have all 46 chromosomes, but for one pair, they came either both from mum or both from dad. The conflict model predicts that if both are from dad—and there is at least one imprinted gene on the chromosome—the offspring should be large. If both are from mum, they should be small. This, and some mutations in imprinted genes or in their regulatory control regions, are associated with a number of human syndromes with odd genetics.

The poster boy for the conflict model is so-called Beckwith-Wiedemann syndrome. This is associated with the chromosome that has *Igf2* and *H19* together, but also a number of other imprinted genes. The babies are born very large and get much larger. They also have a large tongue, which some have interpreted as connected with getting more milk. But they also have distinctive grooves in the earlobes, not an obvious prediction of the conflict model. Nonetheless, the conflict model makes for a good explanation as to why just having two copies of this chromosome from dad leads to large babies—notice we don't need to suggest there is any mutation in the DNA, just a change in dosage of the gene product because of imprinting. Here is a genetic disease without any mutations in the DNA and with the normal count of chromosomes. I told you they were odd.

The same theory also predicts what should happen if you have two maternal copies of this chromosome. We never see these as live births, but there are trisomies (embryos with three copies of one chromosome) with two maternal copies that die in the womb and are, as expected, growth-retarded. A possibly clearer case is so-called Silver-Russell syndrome. This is associated with much-reduced growth in the womb and after. It doesn't seem to be a syndrome with a simple, single genetic cause. Some patients have changes in gene expression of the imprinted genes associated with Beckwith-Wiedemann syndrome. About 5%–10% have two maternal copies of chromosome 7. The maternal UPD cases go in the predicted direction.

UPDs on human chromosome 15 are also interesting. Having two from dad is associated with Angelman syndrome, known in a less sensitive age as happy puppet syndrome. The children are born with a normal birth weight but with a protruding tongue. They tend to end up a little light and short—that doesn't obviously fit the conflict hypothesis. Having two copies from mum results instead in a different condition, Prader-Willi syndrome. As predicted by the conflict model, these babies are born a little small and have difficulties taking mother's milk. All rather as expected of an excess of growth suppressors. What is more enigmatic, however, is that if they do survive, they tend to put on weight because they cannot stop feeding.

For the most part, in mice and humans maternal UPDs are associated with reduced growth. Some of these are only quite recently defined conditions—Temple syndrome, for example, first described in 1991, is a maternal UPD of chromosome 14, and is associated with growth retardation in the womb. This goes with the expectations.

The most unexpected finding is that of the human paternal uniparental disomies—which should result in large babies, as in Beckwith-Wiedemann syndrome—some are associated with severe growth retardation. Human UPD of chromosome 6 is associated with babies that get diabetes just after birth, so-called transient neonatal diabetes. These are at the far end of intrauterine growth retardation, the opposite of what was predicted.

Advocates of the conflict model suggest that UPD data might not be the best data, as such events may be too disruptive. There are, however, other strange facts that don't make a great deal of sense. Many imprinted genes operate in parents in mice and affect maternal care behavior or the mother's milk production. This sort of looks like it might be consistent, but the model says the conflict should all be in the embryos and pups as to how much resource to extract, not in the mother, as she is related in both of her chromosomes to all her progeny. There may be circumstances that could explain this. If mothers feed other mothers' offspring, as mice do, and if the mother can adjust the milk given to the pups depending on whether the pup is hers or not, then conflict theory can predict imprinting in maternal breast tissue. However, the evidence says mums don't adjust their milk production as required.

While one will often see that the conflict hypothesis is taken as if proven, the hypothesis gets a more cautious thumbs up from me. There are a few too many dangling details. But should we even expect one model to get everything right? You might have noticed that in many cases we ask whether one hypothesis can explain all the data, and if not, we dismiss it. For example, in the mutation rate chapter we asked whether the benefit/cost ratio of keeping the mutation rate down can explain all the variation in mutation rates. If it cannot, we tend then to dismiss it. We then ask whether the drift-balance model can do better. It would, however, be odd if the benefit/cost ratio couldn't explain some of what we see. This is a funny thing about my field of study—we seem to presume there to be one explanation even though we know full well that this is frankly ridiculous. However, as a way of stumbling toward the truth it may be an OK way to go.

So too with imprinting. At the last count there were about fifteen different theories for imprinting. Many provide explanations that are so general that they fail to address the observation that most species don't have imprinting. However, there are other potentially important angles to take that might explain some of the data. The other leading argument also supposes that some imprinting may be the result of problems related to coordinating intimate mother-offspring contact. My colleague Jason Wolf noticed that in many cases selection can favor mother–offspring complementarity. Young mice, for example, receive more provisioning if the mother is from the same strain. There are on the market multiple plastic bottles with different-sized screw tops. The bottle works best when the cap and the bottle are of the same "strain."

In humans we also know that your chances of survival are highest as a baby if you have an intermediate birth weight. Big babies and little babies don't tend to survive (they do now with advanced healthcare, but they didn't use to). Wolf's logic then is that there should be a component of the selection on the mother–baby relationship whereby both should coordinate to attain that optimal weight and to generally mesh well—the right bottle top for the right bottle. The conflict model, by contrast, assumes that bigger is better.

This coordination at the mother–placenta interface could be achieved by the placenta expressing mum's version of the gene. Hence imprinting. As we saw before, pre-eclampsia looks to be a problem of the mother mounting an immune attack against her own baby. But if she immunologically "sees" her own proteins, then this should be less likely. Immunologically it could make sense to only express the maternally derived genes in the placenta. For surrogate mothers, this mutual recognition no longer works, and indeed, they have the highest pre-eclampsia rates. Wolf indeed points out one peculiarity, namely that of the imprinted genes that are uniquely expressed in the placenta (i.e., in the placenta and the placenta alone), all are expressed from the maternally inherited version of the gene. There also seem to be more imprinted genes in general expressed from the maternally inherited genes. Others have in turn argued that the activity of genes such as *Zbdf2* fits this model as well, if not better. Its function is to coordinate the newly born pups' needs with the mother, to get her to express milk.

The idea of co-adaption can extend beyond the intimate maternal–fetal interface. For example, in rodents we often see so-called allocare, communal nesting by (usually) related mothers who can nurse someone else's pups. The relatedness of the pups to the nurse is known to influence how well the pups do, as this coordination model would expect. What Wolf's theory predicts is that nurse and pups will express the gene version in which they are more related. The conflict model instead predicts that the expressed gene will be the one for which higher expression leads to greater net benefit of the gene (its inclusive fitness). The test has yet to be done.

Perfection in the Face of Genetic Conflicts—As If

Whether it be mother-baby tussles over sugar, conflicts between genes within the fetus, Machiavellian centromeres, or just higher frequencies of recessive lethals, we have multiple issues getting pregnant, staying pregnant, and having a healthy baby at the end of it. That human reproduction is so mind-bogglingly hard seems to often rest on the shoulders of our unusual continuous provisioning of resources. It could hardly be less perfect.

In many cases we think the problems are underpinned by conflicts of interest: what is best for one is not what is best for the other. To some extent this should come as no surprise. We are very understanding of the notion that what is best for our parasites is not what is best for us. SARS-CoV-2 does well by using us an incubator. Vaccines and face masks are there to stop it from turning us into its incubator.

The oddity we face when thinking about the centromeres or the conflict theory of imprinting is that the disputatious entities are not something *different* from us. They are part of us. The halfway stop may well be the jumping genes and endogenous retroviruses. These are not part of our necessary complement of genes. HIV is a case in point. HIV is a so-called retrovirus. This means it exists as a virus, transmitting from person to person, but it can also integrate into our DNA—that is the retro bit. It goes from being RNA back into DNA and inserts into our chromosomes. If it can do that in a cell that is part of the germline, you now have an insertion with the capability of being part of our species' DNA in perpetuity. It has become endogenous. Like any change to our DNA that can be inherited, this new insertion will start out rare in the population but can increase in frequency. We think the main reason such retroviral insertions increase in frequency is just bobbling about. They don't do any harm so get shoved into our garage.

We see in our DNA waves after waves of similar insertions. Once they get incorporated into our DNA, that can become common and we all then have them. More of our DNA is such endogenous retrovirus than is protein-coding exons and non-coding exons collectively. Eight percent is endogenous retrovirus, 1.2% protein-coding exon, possibly 2.3% long non-coding exon. Don't you find that a strange statistic?

When it comes to endogenous retroviruses and other jumping genes, we have inbuilt responses to try to inactivate them, not least because if they jump back in they could break something important—such jumping genes can be a major source of mutation. They are also likely to affect more babies that never got to be born because a jumping gene landed in the wrong place. We play a cat-and-mouse game with these as well. The defenses that we send to keep them quiet are fast-evolving, as you would expect.

It is perhaps easier to see why we have a conflict with these foreign visitors to our DNA than why we do with what you might think of as resident genes. More generally, the concept of an intra-genomic conflict takes some serious mind-wrapping to get a handle on. I worked on the subject for much of my early career just because I found them so intellectually delicious. They challenge just about every presumption you might have had about evolution and natural selection. I worked on selfish elements that in some insects kill their host if that host is a male (cytoplasmic male killers), others that in certain matings kill all the offspring (cytoplasmic incompatibility) or that destroy half a male's sperm (male meiotic drive) or that kill half the brood in mice (maternal effect lethals) or sterilize pollen production in hermaphroditic plants (cytoplasmic male sterility). Yes, mutations that do all of these things can increase in frequency. They just need to gain by the activity. For example, killing half a male's sperm is pointless if you are in half the ones that die. If, however, you kill the sperm that you are not in, you win. Just take everything you imagine selection could never favor and I suspect I'll find a selfish element that does it.

And do you remember one assumption we made about the canonical process of natural selection that we saw in chapter 3, namely that transmission of mutations is fair and unbiased? The Machiavellian genes upset this particular apple cart and well illustrate why it matters: only when mutations are transmitted in an unbiased manner do their effects on your fitness, not their ability to cheat the system, count.

It would, indeed, be hard to see how evolution by natural selection could occur if transmission from parents to offspring was routinely biased. There would be no concept of perfection. It would be survival of the fittest Machiavellian gene, not survival of the fittest organism. That we do persist is in no small part because most genes most of the time obey the rules of fair inheritance. A bit like a democracy with a rule of law.

Genetic conflicts not only upset the view of a slow progress to perfection, they also can explain some remarkable instances of imperfection. One such happens in mice. The mouse X chromosome partners the Y chromosome in males. The Y is what makes a male a male in mammals. To have a 50:50 sex ratio, males must make as many X-bearing sperm as Y-bearing sperm. But a 50:50 sex ratio is not necessarily in the "interests" of genes on the X chromosome. They are not on the Y chromosomes, so gain nothing from the generation of sons. Mice have on the X chromosome multiple copies of gene *Slx* (about 50 copies), derived quite recently—about 3 million years ago—from another gene, *Sycp3*, that isn't on a sex chromosome. In fact, there are two families, *Slx* and *Slx1*, but we shall just refer to *Slx*. If a male mouse doesn't have enough copies of *Slx* it will be sterile. On the Y chromosome are multiple copies (about 100) of a sequence-related gene, *Sly*. If a mouse doesn't have enough *Sly* it too will be sterile. Curiously, *Sly* and *Slx* have opposite roles. While *Slx* stimulates gene expression on the sex chromosomes after male meiosis, *Sly* suppresses it.

This situation is thus paradoxical. Rodents made sperm just fine before these two sets of gene copy number increases. Ancestrally they didn't need this complexity, but now they have a system with multiple copies on the X chromosome without which the mouse is sterile, multiple copies on the Y without which the mouse is sterile, and the two sets of genes acting antagonistically. The system gets even more puzzling. The deficits we see when males don't have enough *Slx* go away if the mice also have reduced dosage of *Sly*. We can see why if you have lots of *Slx* you need lots of *Sly*—but why do you need either?

What on earth could explain why the mice have lots of copies of antagonizing, rather pointless genes? Genetic conflicts can.

We think the story runs a bit like this. When *Slx* first arrives on the X chromosome it has an advantage, not because it makes for a brilliant mouse, but because more than 50% of the viable sperm of this male contain this X chromosome. This is known as "meiotic drive." Meiotic drive genes are one sort of selfish element and readily increase in frequency just by getting more than their fair share of inheritance. Like more conventional parasites, meiotic drive genes can be bad for the bearer (not too much but a bit) and still go from rare to common.

But there is a big loser here: the Y chromosome. In comes *Sly*. If this stops *Slx* from doing its job, then you could stop the meiotic drive, even possibly reverse it—more sons than daughters. It now also goes from rare to common because it is working for its own good. If this interaction is dose dependent (which it is), then we expect multiple bouts of one of the two increasing copy number, gaining a transmission advantage by so doing, countered by the other increasing copy number, etc. You end up with a ridiculous situation in which no mouse needs either sets of genes, but if the X has lots of *Slx* the Y must have lots of *Sly* (and vice versa), and they trade shots with each other that cancel out.

What would you call that process? Inspired by the cold war buildup of nuclear weapons in the US and USSR—to the point where the world could be destroyed several times over—we call this an *arms race*. If the

US buys more nuclear weapons, the Soviets must too. If the Soviets buy more, the US must too. And, in the end, if one side has all its weapons taken away and the other unleashes all its weapons, the system blows up. Hence the sterility when the multiple copies of either gene family are taken away.

A key prediction, then, is that increasing dosage of *Slx* should result in a female-biased sex ratio in the kids (the X chromosome is being transmitted more). Conversely, decreased copies could end up with a male making more sons—the X–Y balancing act is now pitched toward the Y chromosome. This is, indeed, just what is seen. The male-biased sex ratio associated with *Slx* deficiency is also rescued by *Sly* deficiency.

In short, all the data says that *Slx* increase in copy number and dose is associated with it being transmitted more to the offspring, and that *Sly* has co-evolved to suppress this. And so you end up with possibly the very epitome of imperfection: a system with both sides firing multiple guns at each other but with the bullets mutually annihilating each other in the air. You are better off having neither side firing at each other.

One interesting corollary of this is that this sort of arms race can lead to problems when mice from one population mate with mice from others that may not have the expansions of *Slx/Sly*. Here it is quite possible that male meiosis will just be blown up. This is indeed seen. This, then, is the first step in the process of speciation whereby the two populations become reproductively separated. Speciation—the making of new species—may in part itself be a consequence of conflicts that lead to imperfections.

The *Sly/Slx* story appears not to be a one-off. In flies a similar system was identified many years ago (*Stellate* on the X, *Suppressor of Stellate* on the Y), seen only in *Drosophila melanogaster*, not its closest related species. Recent work suggests these copy number games on the X and the Y chromosome in fruit flies are extremely common. The evidence that selfish elements, such as meiotic drive genes, explain problems that hybrids have is also now well evidenced. It is a funny thought that the reason there may be many species, not one, is itself a consequence of evolutionarily near inevitable imperfection.

In cases like *Slx/Sly* and in similar cases of genetic conflict, when something bad happens we often see a genetic response: *Sly* suppresses damaging *Slx*, we suppress jumping genes so that very few can now jump, *Bub1* suppresses Machiavellian centromeres, *Igf2* and *Igf2r* cancel out. But what about the other problems that we have? We have a high mutation rate, our splicing is error prone, and we have lots of harmful mutations. Is there anything evolution can do about these? In the next chapter we shall see how we fight back, or at least cope with our other genetic misfortunes.

8

The Evolutionary Fightback

If what we have learned so far is correct, we have a problem with mutations. Not just changes from one base pair to another, but also small deletions and insertions, along with jumping genes and retroviruses clogging up our DNA. These are all just different flavors of mutation, hard-encoded heritable changes to DNA. Each new change will initially be rare in the population but then could go up or down in frequency. Most of these changes to our DNA are either very harmful or just a little bit harmful. We have lots of them, and struggle to rid our genome of the latter group. We end up with lots of rare genetic diseases, often die young, and over evolutionary time accumulate junk in our genomic garage. On top of this, with reproductive compensation lethal mutations are more common than they otherwise would be. Such compensation seems also to underpin our ridiculously high rate of chromosomal errors (which are also known as chromosomal mutations).

But is there nothing we can, evolutionarily speaking, do about all this? We have seen some responses already. At least in mice, *Bub1* has counter-evolved to stop selfish centromeres from having their way. *Igf2* and *Igf2r* seem to play off against each other, as do *Slx* and *Sly*. These seem to be examples of evolutionary cat-and-mouse games between our own genes, rather like the repeated interactions we have with our more conventional parasites, such as those that cause malaria. Human populations have seen, for example, evolution of the Duffy blood group protein in response to malaria.

But what about our problems with mutations and the general problem that our cellular processing seems to be less than perfect? Are there also evolutionary responses that organisms like us can reach for? Perhaps there are.

Exonic Splicing Enhancers to the Rescue

One of our biggest inefficiencies seems to stem from the fact that we have such large introns. As we saw in chapter 2, species with lots of intergenic DNA tend to also have large introns. We put on genomic fat everywhere. This is most likely a consequence of weak selection being unable to clear out the junk from the genomic garage. This bloating has several consequences. First, large introns make both copying DNA (replication) and making RNA (transcription) unnecessarily costly. We are producing RNAs that are over 90% stuff that must be removed, spliced out. This is probably why our genes that need to be highly expressed or very rapidly expressed tend to have shorter introns—the cost is just too much to bear, and so selection can more easily remove insertions that greatly increase the size of introns of such genes. That is simple economics.

The second problem is more subtle. It is that large introns make it harder to splice accurately. With each intron the cell has a difficult computational problem: each RNA is a string of nucleotides, but most of these nucleotides need to be spliced out, the remaining sections joined up again. To extract each intron, we cut at one place in the string of nucleotides and then cut downstream of this, remove the stuff between the cuts, and jam the remainder of the nucleotides back together. We do this for every intron.

But what if you cut in the wrong place? The chances are that things will mess up, just as they would if we spliced raw movie film in the wrong places. With our genes having on average about 10 introns each, there is lots of room for errors. Often the incorrect splicing will result in an mRNA that isn't in the right reading frame. Remember, codons are read in blocks of three, so if the splicing disrupts the normal reading frame, the new mRNA will not code for the desired protein and will likely have a stop codon in the wrong place. Typically, a stop codon in the wrong place triggers *nonsense-mediated decay* (NMD), a cellular garbage collector and recycling center. Or perhaps the system got lucky and didn't cause a shift in the reading frame. However, the new transcript may well be missing the instructions for some key bit of the protein, or may have had some spurious extra bit added. This is probably not good either.

Perhaps you can now see the difficulty with having large introns: they make the process of finding the right place to cut much harder. Indeed, if you take an artificial version of a gene with an intron and insert sequence into the intron, often you make the splicing less accurate. The precious exons are like small islands in a sea of intron. Faulty splicing seems rather common. We see that some "exons" are observed in one species only. These are most likely splicing errors. They tend to be near larger introns. Similarly, if we look at the "exons" that are sometimes seen in our pool of mRNA transcripts but sometimes not, they too tend to be flanked by larger introns. Humans also have an especially large number of rare transcripts that break reading frames and don't have conserved splice sites. This all suggests that our introns, especially the larger ones, are quite a nuisance, one that makes splicing harder. We probably tolerate this so long as we can make enough of the protein that we need, but it is hardly ideal.

The problem of doing something about this issue, however, isn't simple. Sure, with nonsense-mediated decay we can try to clean up the mess after it happens. Indeed, in many species introns seem to have booby-trap stop codons, there to trigger NMD when the process misfires and the sequence that should be intron is retained by accident. This is like the emergency services that clean up after a traffic accident sending the mangled cars to be recycled. But why can't we stop mis-splicing from happening?

The issue here again is the problem of inefficient selection. Imagine an intron has a jumping gene inserted in it, for example. The intron now becomes bigger in the one individual with this new insertion. We see this lots—our introns are riddled with such DNA. This may well then have a small effect: reducing the accuracy of splicing, putting more load on the cellular recycling center because the flanking exons are harder to find. The island (exon) hasn't shrunk but the sea of intron has become that bit larger, making the exon search-and-rescue job that bit harder. However, if the effect is just a small reduction in fitness, then selection cannot do much about the insertion, as we have a small effective population size. The same insertion in a species with a large effective population size would be of a large enough effect to be removed by selection. That was Ohta's lesson. That is why, we think, species with small population sizes have bloated genomes.

And herein lies the problem: if selection is too weak to be able to eliminate this insertion from our population, how could it be strong enough to patch in a gene-centered local solution? It can be strong enough to favor effective systems to handle traffic accidents (systems like NMD) because they are happening to many transcripts of many genes at the same time. But can an individual gene prone to mis-splicing do anything about its high rate of erroneous processing?

One possibility is that the fixation of one insertion is just step 1 of the decline. This same process can repeat. Indeed, we often see repetitive sequences like jumping genes insert into a previously inserted, now decaying jumping gene, or next to it. This too we cannot resist because population sizes are small, selection is weak. The ratchet-like increase in intron size has clicked once more. If this increases in frequency and everyone ends up having it, then the splicing process is yet more inaccurate for all in the population.

Can this ratchet-like decay process carry on indefinitely? At some point the splicing is so bad that selection could step in and say, enough is enough. What does that mean in practice? Here we meet a class of DNA/RNA motifs that has fascinated me for many years. They are called *exonic splicing enhancers*, ESEs for short. These appear to be part of our fightback.

With our large introns and small exons, the challenge is to find the tiny exon island in the sea of sequence. Just as people shipwrecked and abandoned on a desert island light a fire so that passing ships can see the small island from a distance, so too we have exonic splicing enhancers (ESEs). What are these exactly? RNA does not light fires! To see this, we need to think a little about the process by which the cell system can figure out where the ends of an intron are in a series of nucleotides in RNA.

There are such things as "splice sites." These are a few characteristic bases on either side of an exon–intron junction in the RNA. These come in weak and strong versions. A strong one is often good enough on its own to say, "Splicing machinery come here, the site to cut is here." Weak ones can sometimes do that. Yeast largely has small introns and strong splice sites. We largely have large introns and weak splice sites. Yeast doesn't have the problems we do. Yeast has large population sizes. Lucky yeast.

We then must build in extra information to tell the system where to go, where to cut. And this is where my beloved ESEs come in. Toward the ends of our exons, we find runs of nucleotides (about 6 bases long) that are "sticky" in the sense that they attract and glue so-called SR proteins to RNA. These ESEs tend to be rich in A and G (especially A). AGGAGG and AAGAAG, for example, are thought to be super sticky. SR proteins are special proteins that sit on RNA and recruit the splicing machinery. They tell the cell, in effect, that somewhere near here you will find a splice site. These special runs of nucleotides, because they enhance the splicing process, are thus known as exonic splicing enhancers. They are fires lit on the small islands in the sea of junk.

Of all organisms, our genes seem to use ESEs more than others. About 30%–40% of all the nucleotides toward the ends of our exons are ESEs, capable of acting like flypaper for SR proteins. Which exons do you think would be especially likely to have a high density of ESEs? If you guessed that it should be exons next to big introns, you guessed right (fig. 8.1). These are the exons that are hardest to correctly identify.

Generally, we find that ESE density toward exon ends is well predicted within a species by the size of the neighboring intron, and across species also by the average size of introns. Species, like us, with bloated genomes, including large introns, are very reliant on ESEs to help the system identify where to cut the RNA. It is part of the fightback against a large and bloated genome. We don't hold the record for the highest density of ESEs. That is held by an unusual brown alga, *Ectocarpus*, a sort of seaweed. Like us, it too has lots of intronic sequence (fig. 8.2).

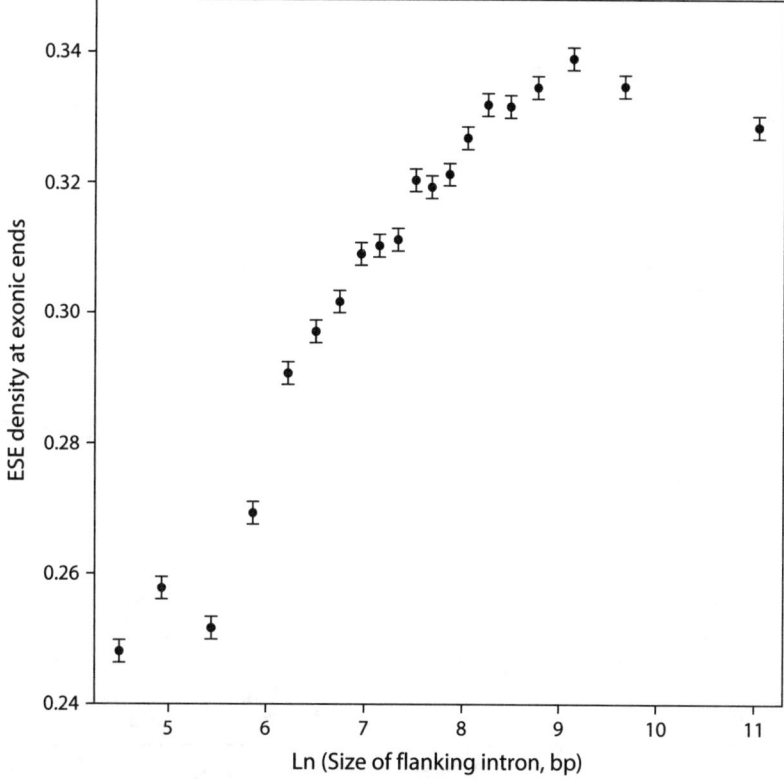

FIGURE 8.1. The density of exonic splicing enhancer (ESE) motifs at the ends of exons as a function of the natural log of the size of the flanking introns.

Our reliance on ESEs to make sure splicing is done properly is in turn evidenced by unusual disease-causing mutations. In all species mutations at splice sites disrupt splicing and cause genetic diseases. Species like us, with a strong reliance on ESEs near to but not at the splice site, are even more vulnerable. Mutations in our DNA that break an ESE when it is in the corresponding RNA have the potential to cause harm. This is because the resulting mRNA will not be properly spliced and so the protein will not be right.

This may indeed be how many mutations that change an amino acid (so-called *missense mutations*) or make a stop codon (*nonsense mutations*) have their disease-inducing activity—they break ESEs, causing

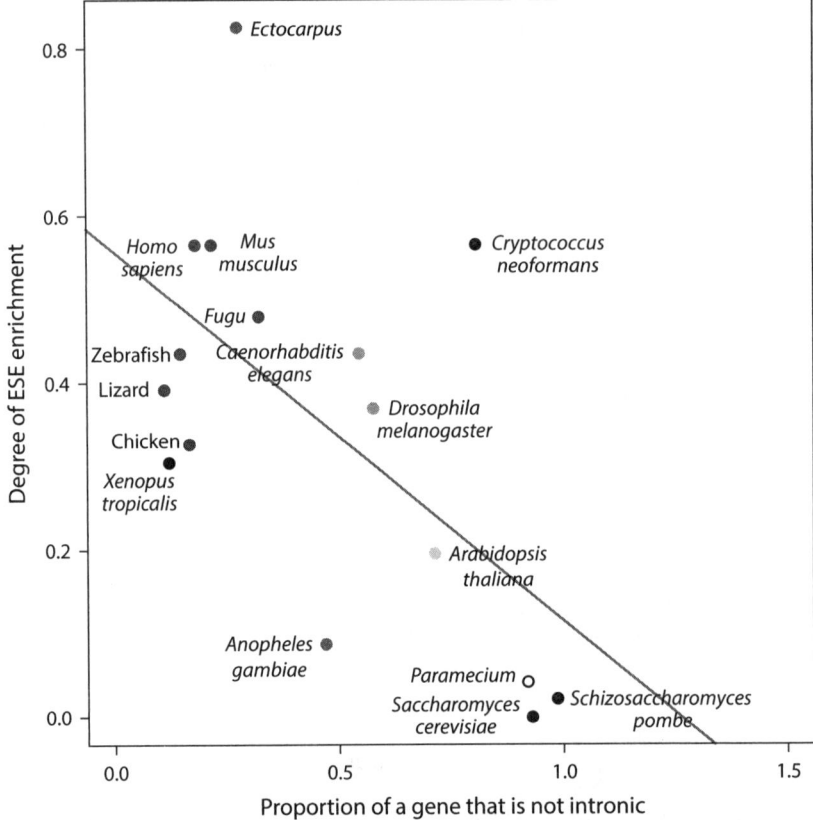

FIGURE 8.2. **ESE enrichment for a variety of model species as a function of proportion of gene sequence that is coding sequence (i.e., not intronic).** Notice that rather little of human genes is protein-coding—they are mostly intronic—and, as in similar species with much intronic sequence, they have lots of ESEs. The ESE enrichment metric here is the proportion of amino acids that showed skewed usage near exon-intron borders. Adapted from Wu, X., et al. 2013. "Evidence for deep phylogenetic conservation of exonic splice-related constraints: splice-related skews at exonic ends in the brown alga *Ectocarpus* are common and resemble those seen in humans." *Genome Biology and Evolution* 5:1731–1745.

splicing defects. ESE mutations may, however, also be *synonymous*, meaning that the mutation converts the sequence to one specifying the same amino acid. For example, one form of hearing loss (called DFNA8/12) is owing to mutations in the *TECTA* gene. The causative mutation is one that converts a CTG codon—specifying the amino acid leucine—to CTA, also specifying leucine. Normally we think of such mutations as "silent"; we presume that they don't matter and must be neutrally evolving, that they just don't affect fitness, cannot cause disease. In this case they do matter. The single synonymous mutation looks to make the RNA much less sticky to one SR protein, SC35. SC35 is a bit unusual: it likes a motif with lots of Cs such as GGCCCCTG. Notice the CTG on the end. The loss of this stickiness leads to a loss of correct splicing—the whole exon is lost from the spliced mRNA. As it happens, this exon was 111 bases long—a multiple of three—so the rest of the protein is correct. However, missing 37 amino acids is enough to cause deafness.

How commonly splicing mutations cause genetic diseases is still under debate. Estimates for the proportion of genetic diseases owing to splicing defects range from around 10% to 50%, but about a quarter to a third seems reasonable. Recent evidence, for example, finds that nonsense mutations (those generating a stop codon) that cause disease are enriched in what were ESEs and are unusually common at the ends of exons, where mutations are more likely to break the splicing process. Indeed, the most recent evidence suggests that, given this enrichment, about 30% of disease-associated nonsense mutations have their effects because they mess up the splicing process rather than because they more directly mess up the protein. We aren't sure what proportion of synonymous mutations have harmful effects when mutated: estimates range from a few percent to perhaps as many as 20%. This remains an active area of research.

What is now clear is that the ends of our exons are odd. They have a high density of ESEs and are especially slow-evolving (mutations happen but are harmful, as they often disrupt splicing). For this reason, you and I are more genetically similar at the ends of our exons than in the center of our exons. Indeed, in our larger exons, the center of our exons evolve about 30% faster than the exon ends. Owing to their strange nucleotide content, ESEs also skew which amino acids are most commonly encoded by the RNA at the ends of our exons. For example, in the last 70 base pairs of our exons (where ESEs reside), we rarely code for proline (CCA, CCT, CCG, CCC) but regularly code for lysine

(AAA or AAG). The nucleotides coding for lysine are sticky for SR proteins; proline's codons for the most part are not. Interestingly, the amino acid arginine can be coded by either sticky AGA and AGG or teflon CGT, CGC, CGG, CGA. Toward the ends of our exons we predominantly use the sticky versions.

This might all seem a little niche: so, there exists a set of short motifs that appear at the ends of our exons that are slow-evolving and if disrupted cause mis-splicing and disease. Why is that important? What cannot be emphasized enough is how odd we are in this regard. The key is that ESEs are common at exon ends, and our exons are short. Remember the 70 bp figure as the zone in which ESEs function. That is 70 bp from either exon end. So, an exon that is 140 bp or less is all exon end. One oddity is that most human exons are shorter than this. With 30%–40% of exon end being ESE, it turns out that a simple metric like the proportion of an mRNA whose sequence is within 70 bp of an exon end is one of only two good predictors of how fast our proteins evolve. The other is how much the protein is expressed. For us, but not for species with fewer and smaller introns, the need to reinforce the error-prone splicing process is possibly one of the most important determinants of how our genes and proteins evolve. In turn, it explains a lot about why mutations kill us or make us ill.

Biased Repair as a Replacement for Biased Selection

Our fightback against error-prone splicing is in many regards an odd solution. It is a genomically local solution to a genomically local problem: the problem any gene has in being spliced is solved by evolution within that same gene. For the most part this sort of solution isn't available: if the harmful mutations cannot be resisted by selection, then selection cannot also favor local compensation. The oddity of splicing is that the buildup of harmful effects can be gradual—a ratcheting of mutations that degrade the ability to splice. More generally, we expect what you might call global solutions. The mutation rate, for example, is owing to changes in repair enzymes that don't just act on DNA in their vicinity, but across the whole of the genome. Because one mutation in a DNA repair enzyme can affect so much DNA, the selection can be much stronger than on genomically local solutions. NMD also acts on mis-spliced transcripts from every gene.

Might there be genomically global solutions to our problems with mutations? One possibility is a funny property of DNA mismatch repair.

DNA repair generally is all about resisting mutations by stopping them from happening. Every time we get a mismatch between the two DNA strands, mismatch repair is there to try to stop mutation from happening. If we started out with a GC pair in the DNA, G on one strand, its complement C on the other, something could go wrong, and we could get an AG mismatch, for example. A and G are not supposed to pair up. G and C pair up as do A and T, but no other combinations are permitted. If not corrected, we could have ourselves a new mutation. But if we can resolve the AG mismatch, take away the A, replace with a C, then, phew, we have stopped a mutation from happening.

But there is a problem. How did the system know that the G in the AG mismatch was the correct base? If the system resolved the mismatch by going with the A, then, in this case, we have made a mutation by re-pair: a GC pair went via a mismatch to an AT pair. In a sense the mismatch repair has done its job—there is no mismatch, as we now have an AT pairing. But we now have a different problem, a new mutation.

If we cannot know which base might be the original one, could there be some rule of thumb that we could apply instead that keeps the selectively better nucleotide more often than not? We think that there might be.

To see this, think about a barrel full of black or white balls. Each black ball has a probability of spontaneously becoming a white ball, each white ball has a chance of flipping to be a black ball. The resemblance to mutation I hope is obvious. Let's follow this barrel of balls over time. At each time point I'll count the number of white and black balls and work out the ratio. When after some time the proportion of the white to black balls stops changing, I'll conclude that we are at the balance point, the equilibrium. At this equilibrium things don't just stop, how-ever. Rather, for every white ball turning black there is a black ball turn-ing white. This is known as a dynamic equilibrium.

What, then, is the relationship between the equilibrium ratio of black to white and the relative rates at which black changes to white and vice versa? If the rate at which white flips to black is the same as the rate that black flips to white, the equilibrium proportion will be 50:50. But what if, for example, blacks flip to white more often than white flips to black? Fifty-fifty cannot now be the equilibrium. If we had a barrel with 50:50 black to white and came back a little later, there should be more whites than blacks, just because black balls are more mutationally unstable. We weren't therefore at any balance point.

Perhaps your intuition is that if blacks flip to white more often than the reverse, then the barrel should go to a state where there are only white balls. Your intuition isn't far off. You would, however, be wrong for an interesting reason. As the blacks become ever rarer and whites ever more common, the net chance of a white flipping to a black goes up simply because there are more whites. The net rate of flipping black to white is the frequency of blacks times the chance the black-to-white flip occurs. As the blacks get rarer the black-to-white rate goes down. Indeed, think about what would happen if there were only whites, for example. In this case, all mutations would be white to black as there were no black balls. So, all white is not an equilibrium either.

The equilibrium position is in fact just a property of the relative rates of flipping white to black and black to white. There will be a frequency between 50:50 and 100:0 where the rate at which a new black ball is created (frequency of white balls times rate of flipping white to black) is exactly equal to the rate at which a new white ball is created (frequency of black balls times rate of flipping black to white). The bigger the relative difference in flipping rate, the more extreme the equilibrium bias.

Mutation is like this. Mutations are random in the sense that where they hit in the DNA and the effects that they have are not directed. We looked at that in chapter 6. But the process is biased. For us, as with many organisms, it is more likely that a G or C will mutate to an A or T than the reverse. This is a mutation bias, but in outline it is the same as our black-white flipping. A consequence of this is that if the only evolutionary process affecting G+C composition is mutation (with a mutational bias), then DNA would evolve to a position (mutational equilibrium) where As and Ts are more common than Gs and Cs, just as the black/white ratio is affected by a flipping bias. In our DNA, from what we know about the mutational process—by studying DNA of parents and their offspring—we estimate that about 28% G+C is the mutational equilibrium. At this point, for every G or C created by mutation from an A or T somewhere in our DNA there is an equal but opposite A or T created from a G or C somewhere else.

The assumption that nothing else is acting on G+C content is equivalent to assuming that new mutations are subject to drift (neutral evolution) alone. What would you then think if you studied a gene and found that it was much more than 28% G+C? At this point you should be scratching your head. If the only process determining G+C is mutation bias and this predicts 28% G+C, we need to suppose that our simple model is missing something. The most obvious thing it could be missing is selection. Genes code for proteins, and proteins will often need codons that are G+C rich to specify the amino acids that you need. Glycine is, for example, a common and important amino acid. But all glycine codons start GG. Thus, if we see a gene with much more than 28% G+C content, you could fairly suppose that the excess Gs and Cs, beyond the mutational equilibrium, are there for some good reason.

Now let me give you that mismatch event again. We have a mismatch between A and G. We don't know which was the original base and which the wrong one. But we aren't interested necessarily in which was the original—we are more concerned with which is the better one, if there is a better one. Perhaps you can see the rule of thumb to use? Because more of the time we need a special explanation for Gs and Cs, as mutation bias is in the opposite direction, you should suppose that the G+C is, on the balance of probabilities, the base more likely to be beneficial. What will you now do when you meet an AG mismatch (or a TG or AC or TC mismatch)? The sensible thing to do is preserve the G or C and patch in appropriately.

Remarkably, we use a system like this to mend mutations that have already happened. When we make our sex cells (sperm and eggs), we exchange DNA during the crossing-over process that we touched on in the last chapter. Part of this process is a bit involved: it involves one strand from the chromosome you inherited from one of your parents (let's say mum) invading the double-strand DNA at the same position on the other chromosome (in this case, the DNA you inherited from your dad). You now have a weird structure—a stretch of DNA with one strand from mum, the partner strand from dad. This is called a *hetero-duplex*, *hetero* meaning "mixed," "duplex" referring to the double-stranded nature of this structure. But what if your dad's DNA at one position had an AT pair while you mum at that same position had a GC? In the heteroduplex you could well now have an AT GC mismatch, perhaps A on one strand, G on the other. Mismatch repair kicks in. But what is remarkable is that it doesn't flip a coin to decide which to go with. About 70% of the time it goes with the G or C. This is known as *GC-biased gene conversion*, or gBGC for short. Biased gene conversion is a hot topic in evolutionary genomics presently.

What effect will gBGC have on the way our genome evolves? It has some remarkable properties. The net effect of gBGC will depend on the length of the heteroduplex, how often they form, the frequency of AT–GC mismatches, and the bias, which as I mentioned seems to be about 70:30 in favor of G and C. Let's start by thinking about a site in our DNA where it doesn't matter which nucleotide you have, any one of the four is fine. In this case, gBGC has the effect of increasing the frequency of Gs and Cs simply because of the 70:30 bias. This bias in somewhat like the bias we saw for selfish centromeres. Simply by getting into more than 50% of the eggs or sperm, a mutation can go from rare to common.

We therefore expect to see a few things. First, where in our DNA recombination is more common there should be more Gs and Cs just because heteroduplexes are more likely to happen there, so gBGC should be more common. We indeed see this. Our DNA is in fact odd in that it has sections with very high G+C in the DNA between genes and in introns, and other sections with much lower G+C content. These G+C-rich or A+T-rich blocks are over 300 kb long and are called *iso-chores* (they aren't really all that block-like, more wavelike, but we needn't go there). The high G+C regions do lots of recombination— for example, the tips of our chromosomes. The A+T-rich regions are the opposite. Remember that our centromeres don't do recombination—

they are A+T rich. Our Y chromosome is especially interesting. At its tips it will exchange DNA with the X chromosome. Otherwise, it has no partner to recombine with. Those tips have high G+C, the rest of the DNA is A+T rich.

The net effect of gBGC is to force a G+C content well above the mutational equilibrium. In effect, if selection cannot preserve beneficial Gs and Cs for us, our repair process is tuned to do it for us. In this sense, gBGC looks like it is a genomically global system that on the average stops deleterious mutations (commonly GC to AT) from spreading just because selection is too weak to stop them.

How important gBGC is outside of humans is not so clear. It seems to be important in mammals and birds. However, in baker's yeast, for example, we have attempted to estimate whether there is any bias. Yeast is great for this, as you can study all the four products of one meiosis— this is known as *tetrad* analysis. We estimated that, rather than a 70:30 bias, the effect seen in 100,000 mismatches was about 50.03 to 49.97. Almost no bias. Whether this bias is real we don't know, as we are close to the limits at which the techniques are accurate enough. It seems clear that there isn't a 70:30 bias.

This tempts the question as to why birds and mammals have strong gBGC and yeast does not. We really don't know the answer to this one. One possibility, however, is that it all comes back down to the efficiency of selection. Selection is good at getting rid of deleterious mutations in yeast, as it has a large effective population size. Thus, mutations that reach appreciable frequencies in the population are unlikely to be significantly deleterious; selection would have weeded them out first. In us and birds, with small populations and inefficient selection, it is quite likely that harmful mutations (often GC to AT) would rise to appreciable frequencies. Letting gBGC get rid of them for us, as selection is weak, is then possibly a good genomically global strategy. If this idea is right, it suggests that the net strength of gBGC is tuned to the probability that a heteroduplex is made between a good and a bad version of the same bit of DNA. Small populations make selection less effective, so make this more likely. This suggestion, however, has yet to be tested. We need to know more about gBGC and its magnitude. With new sequencing technologies this is now possible. We just need as many estimates for gBGC's effects as we have measures of the mutation rate and effective population size.

With gBGC We Can Shoot Ourselves in the Foot

One thing we can be more confident about is that, as the GC preference is just a rule of thumb, it will sometimes get it wrong and promote harmful mutations. Recently, a clear example of this has been discovered. To be confident that gBGC is increasing the frequency in the population of harmful AT to GC mutations, we need to have a case where we can be pretty confident that A or T is good, but G or C is bad. Stop codons seem to provide just this circumstance.

If you recall, there are three different codons that say "STOP translating" to the ribosome. These are TAA, TGA, and TAG. What is unusual about these three is that TAA seems to be universally the best one. By "best one" we simply mean that the ribosome is less likely to misread the stop signal. TAA is missed by the ribosome about one in every 2,000 times; TGA by contrast is misread one in 500 times. Such readthrough usually leads to all sorts of problems. It could add a bit of protein on to the end of a protein, which is expensive; it can lead to misfolding or to the protein going to the wrong place in the cell. Alternatively, the ribosome can go crashing into the runs of As at the ends of mRNAs. This run of As seems to be a sand trap for out-of-control ribosomes. Recovery vehicles are then needed for the ribosome, the protein, and the RNA. A real mess. Not surprisingly, we see in most organisms evidence that selection acts to reduce readthrough rates. For example, after the stop codon, the very next base modulates readthrough rates. The bases associated with higher readthrough rates are avoided.

TAA is then thought to be the best one to use. Indeed, when we look at highly expressed genes—where cellular traffic accidents are the most damaging—in all species, from bacteria through to humans, there is a preference for TAA above what you would expect from nucleotide content alone. In nearly all other species (bacteria, invertebrates, etc.), we also see that TAA is conserved and preferred. Not in humans. We conserve TGA. You read that right—in us the least good, most leaky stop codon is conserved. We have, consequently, high rates of cellular traffic accidents.

And it isn't only us. We discovered that the extent to which TGA is conserved is predicted by the extent to which gBGC operates. Species with strongly isochoric genomes like ours, with blocks of high G+C and other blocks of low G+C, are those most prone to overusing and conserving TGA. This makes sense if TAA to TGA mutations are harmful but being promoted by gBGC. Indeed, as predicted, in our genome TGA is conserved in the domains where recombination is most common. TGA is also overused more on our shorter chromosomes, as they have proportionately more recombination per unit length.

Why then do we conserve TGA? Not because it is a brilliant stop codon, not because selection favors readthrough and traffic accidents. It seems to be that we have a rule that when faced with a mismatch, go with the G and not the A. As a consequence, when in a heteroduplex TAA and TGA pair up (or rather, TAA with TGA's opposite strand, ACT), we go with the GC. As expected, this is nothing to do with being a stop codon: the same happens in introns, in intergenic sequence—everywhere. And its isn't just TGA and TAA doing this. Many of our amino acids are coded by codons that end in either A or G, on the one hand, or C or T, on the other. In our genomic domains with higher recombination rates, we conserve the G- and C-ending ones, not because they are good, necessarily, rather because that is a consequence of following the GC-over-AT rule of thumb. In the parts of our DNA that don't recombine much, we seem to be, outside of genes, very close to mutational equilibrium.

Biased gene conversion breaks one of our hidden conditions about how adaptation works: that a mutation in an individual that has one old, one new version of the DNA transmits each with equal probability to the offspring. As we have seen, if this is true, we have a process that can drive a bad mutation to higher frequencies (hence we use too much TGA). In the case of biased gene conversion, the biasing of the transmission is something we do to the DNA, rather than something the DNA does on its own.

Is Sex Part of the Fightback?

While GC-biased gene conversion has obvious and immediate effects on DNA, perhaps the more profound corrective to decaying genomes when population sizes are low is the more subtle effect of sex and the genetic mixing that this entails.

For humans and other mammals, the fact that we reproduce by sex, meaning our offspring are a combination of the DNA of our mum and dad, is not a mystery. We have no choice. As we saw in chapter 7, we have a system of genomic imprinting in which, during our development in the womb, some important genes are expressed exclusively from the DNA we inherited from our fathers. An unfertilized egg that somehow starts to develop will rapidly run into trouble.

Birds have a different problem. While in humans and most other mammals, the male has the pair of dissimilar sex chromosomes (X and Y) and the female has two Xs, in birds it is the other way around. To not confuse the two systems, we refer to the bird system as ZW sex determination. A female bird has the two different sex chromosomes (Z and W) while the male has the two similar ones (ZZ). Imagine a female who makes eggs, with one copy of the chromosomes, that can then spontaneously start to develop without being fertilized. This usually needs a first step wherein the egg with one copy of all chromosomes just doubles these up, so now resembling a fertilized egg. The embryos will all then be either WW or ZZ. WW embryos go nowhere. ZZs can develop. There are indeed rare reports in turkeys, for example, of embryos spontaneously developing like this. Even if they were perfectly fit and healthy (they tend not to be), they are all males! And so after just one generation, an asexual lineage grinds to a halt.

This block need not always hold. If a bird were to double up all the chromosomes by chance before making eggs, then the eggs would already have two copies and some could be ZW. However, asexual reproduction in birds, while occasionally reported, seems to be a weird one-off event, not a sustainable strategy. We know of a few cases from species held in zoos—here we can be confident that a female really has not mated, and it is often assumed that just once in a while, if an egg cannot be fertilized, it might spontaneously start to develop. Recently a strange case was reported from the Beckman Center for Conservation Research of the San Diego Zoo in California, of two female condors held in a captive breeding colony. Both mothers had a ZZ son that were shown genetically to have only a mother. This is odd in itself. Odder still, a male was held with them who had fathered other kids.

With the exceptions of birds and mammals, we do see, in many groups of organisms, species in which females routinely reproduce without any genetic input from males (not just a few weird occasional oddities). There are about 50 species of fish, amphibians, or reptiles that have no males and in which females make daughters without fathers. New Mexico whiptail lizards and Amazon molly fish, for example, make eggs that develop without the need of a genetic contribution from sperm. Oddly, in the molly, the females need to have sex with males of a related species, the sperm being used to trigger the development of the egg but not making any genetic contribution to the offspring.

Many wasps also are regularly asexual. The parasitic wasp *Lysiphlebus fabarum*, for example, has individuals that are asexual (mothers make daughters without fathers) and others that are sexual. In this case it looks like a female's "choice" to be asexual is dependent on one gene alone. A female needs both of its copies of a certain gene to be one of two varieties to be asexual. You may well ask how, if the females are asexual, could we know this? Surely we would need to mate male and female to understand the genetics, but, by definition, these females don't mate with males. It turns out that about one in every 3,000 offspring of the asexual females are males. These rare males can then be employed to mate with the sexual females to discover the genetics of asexuality.

In not all cases does asexuality go via an egg. Starfishes, for example can break off a limb. This one arm will grow to a full starfish and the "parental" starfish will regrow the missing limb. This is still one individual becoming two. In *Hydra*, a tube-like invertebrate with a crown of arms, a brand-new small but fully formed baby hydra is generated by budding off from the parent.

One of my favorite similar examples is a recently successful plant in the UK. *Veronica filiformis* is a small blue-petaled plant that you will see on lawns through much of the UK (commonly called a speedwell). It isn't a native plant. It is normally found in Turkey and the Caucasus, but was introduced into the UK in 1808 as a cultivated plant. However, it has escaped from gardens, being first reported in the wild in 1838. More recently it has spread more dramatically to become a perennial feature on UK lawns. All the plants in the UK are, however, sterile (I often look when I see one to double-check—I've never found a seed-bearing one). How could a sterile plant be so successful?

What is unusual about this plant is that it spreads by lawnmower and raking. Every time a lawnmower runs over the plants, the cut fragments can regrow, making multiple new plants. The plant is spreading into Western Europe as well. It seems to be that the plant was packed around the roots of vine shoots taken from Georgia (just south of Russia, not the US one) in the later part of the nineteenth century. From there, whenever it gets fragmented it spreads. An analysis of introduced plants around Moscow suggests that every year it spreads 4 km north and 10 km south. That is pretty impressive for a plant that can't walk (obviously) and can't make seeds.

Asexuality is usually considered an evolutionary conundrum. All asexuality looks, from an evolutionary point of view, like a really good idea. You (mostly) don't need males, so aren't restricted by having to find a partner. Unless you are a species with ZW sex chromosomes, you don't make sons, and sons are, as we saw before, a bit useless. A mother needs to invest energy in making sons, but sons don't then invest in the next generation. Their contribution is one tiny sperm and its DNA. If the sex ratio is 50:50, as we saw is often the case, then an asexual lineage will be able to expand and take over fast. It can go from one female to 2 to 4 to 8 to 16 to 32 to 64, etc. Meanwhile, the same female, if sexual, will on average leave one son and one daughter as surviving progeny and so see no population growth. *Veronica*'s rapid invasion seems to be testament to the efficiency of not needing males.

However, we also see that when we find asexual species, they are an isolated evolutionary twig on the tree of life, the surrounding twigs all being sexual. The nearest relatives of *Veronica filiformis* are sexual speedwells, of which there are many different species. What this suggests is that in the longer term, asexuality as a strategy is doomed. If it was a great long-term strategy, most things would be asexual and the sexual species the weird oddities, the funny little twigs on the evolutionary tree.

There may be some possible strange exceptions to this rule that asexual species are twigs, not bushy branches on the tree of life. The most prominent of these are the several hundred species of bdelloid rotifers (pronounced "delloid," the b is silent). These are tiny, millimeter-long tubular animals in which no male has ever been seen. Have a deep look in moss or in water trapped in your gutters and you may well find some of these intriguing invertebrates. A species-rich group that are all asexual would be mighty unusual. If this is the case, they all haven't had sex for over 40 million years (if ever you think your love life is lousy, just contemplate the bdelloid). As this is so enigmatic, John Maynard Smith labeled them an "evolutionary scandal." However, more recent genetic evidence finds that they have the genetic hallmarks of a species that is sexual. Perhaps they are less scandalous than we thought.

It is also thought that in the short term sexual reproduction must have some advantages. In a few species we see both sexual and asexual members of the same species coexisting. If asexuality was always even in the short term such a great idea, they should take over completely. But they don't.

In New Zealand's lakes there is one of the best-studied instances, a species of snail, *Potamopyrgus antipodarum*, that has both sexual and asexual representatives. Over a decade, it was found that the relative numbers fluctuate, but the asexuals didn't take over. The sexuals must be doing something right.

What are these short-term advantages to being sexual? While easily asked, it turns out that this is one of the most profound unresolved questions in biology. At the last count there were well over twenty theories, not all of which need be mutually incompatible. Indeed, many suggest that the problem of sex is unresolved because the answer probably is not one explanation but an interplay of many different forces. One likely part of the mix is that sex can really help purge harmful mutations from populations. In this sense it is part of the fightback, or at least part of the answer as to how we can still be here despite the inefficiency of selection.

There are several different mechanisms by which sex is good against harmful mutations, but one is especially important to us, as it is most influential when populations are small. Consider this, for example. Imagine we have a species reproducing asexually. We can discern the DNA sequence of each individual and from this, calculate both what the mutation rate is and how many harmful mutations each individual has. A lucky few will have very few mutations; some will have lots. Those with lots are more likely to die (by definition), so selection will favor those with few mutations. But mutation will also tend to result in individuals and their offspring increasing the load of harmful DNA. The two forces, selection and mutation, go in opposite directions. We can imagine that the population could then come to some sort of equilibrium, mutation increasing the burden, selection reducing it.

But now let's also suppose that this is not an especially large population. If so, then the number of individuals with the lowest number of harmful mutations may be small. What if, by chance, all these individuals died early or otherwise didn't get to reproduce? The smaller the population, the more likely that that will happen. What then?

What the American geneticist H. J. Muller noticed was that, if this species is asexual, then there is no realistic way that the population can again have any individuals with this low number of mutations. The best state to be in is now a little worse than the prior best. But then we again face the same problem. In a small population, this new best class can all be lost owing to chance. With asexual reproduction there is no easy way to generate individuals with very few mutations. Each time the genetically best class is lost, it is like a ratchet that has clicked one position, never to go back. This process is then called *Muller's ratchet*.

Muller's ratchet should lead to the small asexual population genetically decaying over time. Indeed, the process may accelerate. For each click of the ratchet, the best individuals are less good in absolute terms. If this means that the population also becomes smaller, then it will also become ever more vulnerable to chance loss of the best types. This, it is suggested, could lead to a process of so-called mutational meltdown. Asexual small populations rapidly become smaller populations and genetically decay even faster.

Muller's ongoing ratchet is, however, easy to halt. Just have sex. When sex occurs, individuals can have offspring that have fewer mutations than either parent. That is what sex does: it generates variation. Kids are both different from each other and different from their parents. Sure, some will have loads of mutations, but a lucky few will have very few mutations.

Another mode by which sex might be beneficial to us applies to species that, like us, have unusually high mutation rates. Unlike Muller's ratchet, this explanation does not evoke random processes associated with small populations. Rather, for this model to work we need just two conditions: that the mutation rate is high enough, and that the more harmful mutations you have the worse off you are. And by worse off, I don't just mean worse off because you simply have more harmful mutations. Rather, I mean that each new mutation disproportionately reduces your fitness. If you have two mutations, for example, you need to be worse off than you would have expected given how badly you do with having just one bad mutation.

An extreme version of this sort of effect—and one that makes explaining the model easier—is called *truncation selection*. Under truncation selection, an organism with one mutation is fine. One with two is fine. But at some point, one extra mutation and you are dead. You just fell off a mutational cliff. Nothing actually has truncation selection, but it is an extreme form of the sorts of interactions we are thinking about.

Imagine then an asexual species in a world with truncation selection. Such a population will evolve such that all individuals will sit right at that cliff edge. If having 10 mutations you are fine but having the 11th kills you, then the individuals alive today will have 10 mutations. Each individual that has a further mutation, however, is dead (I said it wasn't especially realistic). In this species the death rate is the mutation rate. The higher the mutation rate, the more the asexuals do badly.

What about a sexual species? Will it do any better? The answer can be a profound yes. Each time individuals mate, some offspring will have few harmful mutations, but, importantly, some babies will have loads of harmful mutations, many more than the 11 needed to kill you. This makes the population as a whole rather efficient and provides a direct benefit to the mothers to reproduce sexually. Sex could create a baby with, perhaps, 15 mutations (notice no asexual could ever have 15 mutations if the cliff is at 11). The death of an individual with, say, 15 mutations has purged many mutations from the population. One death, 5 excess mutations have gone. Each death of an asexual purges only one mutation.

We can then calculate just how high a mutation rate is needed for a sexual species to overcome the large costs otherwise associated with being sexual. Rather amazingly, after pages of horribly complicated math, we come up with a simple answer. One. Yes, that is right. If the species has one or more harmful mutations per generation per genome, then that species is better off being sexual than asexual.

Such was the beauty of this prediction that it triggered a lot of interested parties to try to estimate just what that number of new deleterious mutations is for any species. It also stimulated a lot of work asking how the harmful effects of mutations multiply when you have ever-increasing numbers of them. For us, with our remarkably high mutation rate, we probably do (just) have more than one deleterious mutation per genome per generation. We are, however, unusual. Our lab fruit fly only ever reproduces sexually, but its deleterious rate is probably a bit under 1. For most species with big populations and short generation times, the number is probably a bit under 1. The studies of how mutations interact were also not especially encouraging. Certainly, some harmful mutations interact as the model requires. But many do not.

Owing to these results, this model is not now thought to be the complete explanation. However, whether it is part of the explanation is less clear. The mutation rate needed is only needed if this model is the only advantage to sexual reproduction.

Conversely, Muller's ratchet model has survived testing rather better. For example, one important study looked at the bacterium *Salmonella typhimurium*. The experiment started with 444 lineages, each allowed to grow on an agar plate, and then, from each plate one bacterium was randomly selected to propagate to the next plate. The researchers did this grow, pick, transfer process once per day for all 444 lineages for 60 days. Each day there were about 28 cell divisions from the initial bacterium that was picked, making about 1,700 cell generations. The conditions had two effects. First, the experiment forced the bacteria to be asexual. Second, picking just one bacterium each time, allowing the population to regrow, then again picking just one, forces strong random loss of different types, the process that underpins the loss of the most fit in Muller's ratchet.

What they observed went with Muller's expectations. Of the 444 lineages, 5 strains were much slower-growing at the end. This bacterium takes 23.2 minutes to do one cell division, but in these five, that had slowed to between 25 and 47.5 minutes. No lineage had increased its growth rate. It was concluded that, indeed, small populations had forced the populations to genomically decay.

A subtly different experiment has been done in baker's yeast, *Saccharomyces cerevisiae*. In this case, each day the researchers transferred a fixed dilution from one tube to another. The fixed dilution ensured that few cells were transferred, but also that the number of individuals transferred from day to day went down if the yeast were not good at growing. If mutational meltdown is a real process, this setup could find it. They did 175 daily transfers, about 2,900 cell divisions. As the theory also says that Muller's ratchet should be more of a problem if the mutation rate is high, they also started to replicate lines either with a strain that had a high mutation rate or a strain with the normal low mutation rate.

For 12 starting lines they found that two went extinct, both of which had the high mutation rate. At least for one of these, the extinction looks to be owing to a recent major reduction in fitness owing to sampling of so few individuals. As the researchers froze down the samples as they went along (to analyze later), they could also see if this extinction was repeatable. Indeed, it was: every time they repeated the experiment starting from the population that just had the fitness decline, the populations always went extinct. If, however, they started from a population that was taken from before the sudden decline, they didn't see the extinction. Not only does Muller's ratchet lead to fitness decline, but if population size goes down with population fitness, then fitness can spiral down.

Muller's ratchet is by no means the only problem of being asexual. Another issue that compounds the issues is that by not having recombination and exchange of DNA, if a good mutation arises, this can increase in frequency and spread so that everyone in the population has it somewhere down the line. That is a good thing. But with no re-combination, all those descendants also have all the other mutations that were present in the original asexual genome. In effect, everyone is now genetically identical. This is the "hitchhiking" we met in chapter 6. All being identical, even if there are lots of individuals there is only really one, genetically speaking. The effective population size has crashed. As we have seen, if the effective population size is low, slightly harmful mutations will accumulate.

Such experiments are all well and good, but does sex explain, in part, why we survive and cope with deleterious mutations? Here nature has given us a rather lovely sort of natural experiment. The Y chromosome is a large chromosome that, except at its very tips, doesn't have any DNA exchange. All other chromosomes have crossing over, the exchange of DNA between mum's chromosomes and dad's when we all make sex cells. The Y chromosome is like an asexual genome in a sexual species.

If sex and crossing over were important to keeping deleterious muta-tions at bay, the Y chromosome should then be the odd one out. It should show hallmarks of genetic decay. And indeed it does. There are now very few genes on the Y, but we can see traces of old ones that gained mutations that stopped them from functioning, such as prema-ture stop codons or insertions and deletions.

It is even more valuable to try and see this process in action. In *Drosophila miranda* such an opportunity presents itself. This fruit fly has a very young new section of the Y chromosome (being new we call it a neo-Y). It looks to be less than 2 million years old. As expected, this neo-Y has very little variation on it, and many of the genes left on it have major disruptive mutations. The faster rate of protein evolution and greater divergence from the closest related species of the genes on the neo-Y also suggests that selection is less effective.

Across our genome we also see fingerprints of what is expected if crossing over (genetic jumbling-up when making sex cells) is saving us from genetic decline. For example, in the parts of our DNA that recombine more we see more variation, as there is less influence of hitchhiking. We are all most genetically similar at our non-recombining centromeres and most different from each other at the highly recombining tips of our chromosomes.

Sex, then, is a solution to mutational decay, especially in species with small effective population sizes. This is almost certainly not the only reason sex is an advantage. Generally, sex shuffles genomes, and a consequence of this is that beneficial mutations can escape the genomic context in which they first arrived. This shuffling in effect allows a beneficial mutation to, on average, sit in a random background and so get fairly judged by selection against different versions of the same gene, rather being judged by the effects of the neighbors it first moved next to. If a beneficial mutation first arrives in a poor-quality genome in an asexual by contrast, it will always be in the same rather bad genome and has little chance of going anywhere, despite being beneficial.

This sort of process is thought to underpin the other great advantage sex has for us, namely that it allows us to keep up with our parasites. A problem with being large-bodied and long-lived is that our parasites can have many more generations for every one of ours. As it takes twenty years or so for one human generation, we evolve so much more slowly than our faster-reproducing foes. Consequently, mutations in our genome that help keep parasites at bay need every helping hand they can get. Not getting bogged down with lousy neighbors and not having all the kids genetically identical are probably two important components of how we, as a species, haven't yet been eliminated by parasites. Indeed, this sort of process seems to explain why the New Zealand sexual snails were not outcompeted by their asexual counterparts. When an asexual clone gets to be common, they get infected with small parasites. In turn, a new asexual clone can come in, but it too eventually gets whacked. If you are an asexual clone, it only takes one parasite that evolves to be able to infect one of you to infect all of you. Throughout these comings and goings, the sexual species persist with lower levels of the parasite. A parasite that can kill one of your kids may not be able to kill all of them, as they are genetically different.

These ideas are more than theory. Experiments using a bacterium that attacks lab worms finds that forcing the worms to co-evolve with the bacteria leads to increased rates of sexual reproduction, while those populations with individuals who could only fertilize themselves went extinct.

Whether it be exonic splicing enhancers to reset our splicing, biased gene conversion to try to do selection's job for it, or sexual reproduction to help rid us of harmful mutations, it looks like, despite having problems with mutations, we get by, evolutionarily speaking. Just. You do, however, have one most unusual asset, not open to other species. You have a big brain. In the last chapter I will consider how that, more than anything, allows us to fight back against our rather rotten evolutionary lot. Medicine is our best anti-mutation weapon.

9

The Medical Fightback

It looks like humans are anything but the pinnacle of evolution. Put differently, I am mostly genomic junk getting by. Seen from this new view, evolution by natural selection and constant progress are far from inevitable. The climb to perfection needs large populations, low mutation rates, unbiased transmission of mutations, and particular ways mutations interact. We evolved with small populations, high mutation rates, and a mode of reproduction that, while seeming lovely—nurturing our young in the womb—leaves us vulnerable to biased transmission of mutations and genetic conflicts of many flavors.

We have seen how evolution has enabled some sort of fightback. Sexual reproduction helps remove deleterious mutations, at least in the longer term. ESEs help lessen the burden of splicing errors, and biased gene conversion, while giving us poor stop codons, probably helps keep our genes healthy. We have evolved genetic resistance to some diseases (only for the disease-causing species to in turn evolve again).

But we humans have an even better trick. With big brains, we have come up with medicine.

Medicine is our great hope in the fightback against the tyranny of mutation. More generally, medicine is our answer to natural selection. Many of us reading this would, in another age, be dead by now, part of natural selection's grim cull. Many of us reading this would, if born in another part of the modern world (or a different neighborhood), be dead by now for the same reason. Medicine keeps people alive blunting natural selection's ruthless sword.

We need not look so far back in time to see this. British data from the 1930s and 1940s was compiled to examine the survival prospects (to 28 days) of babies of differing birth weights. There was an optimal weight at about 8 pounds (3.6 kilos). Around 2% of babies at this weight died before 28 days. Deviate a little from this and your chances were much worse. At 5.5 pounds (2.5 kilos), what is now considered "low birth weight," 10% were dead by 28 days. Half of babies that were 3.5 lbs (1.6 kilos) died young ("very low" birth weight is defined as less than 1.5 kilos). By contrast, if we analyze data from the 1950s to 1979, we see a steady reduction in 28-day mortality rate over time, with about 1%–2% of babies around 2.5 kilos dying within 28 days by 1979. The reduced mortality rate is even more pronounced for the very smallest babies. This can in no small part be attributed to changes in the care of babies, such as policies on the use of oxygen introduced in 1950s. Presently in the UK the neonatal 28-day mortality rate is around 2%, the same as the rate of those of the optimal weight in 1930s.

This increased survival need not, however, have any effects on the activity of natural selection. Selection requires that at least some of the variation between babies in their birth weight had a genetic component. If there is no genetic involvement, then death, or lack thereof, will have no effects on the genetic constitution of this or following generations. For example, if small babies are delivered by mothers who simply didn't have enough food, then the variation in birth weight would be directly owing to environmental causes alone. For birth weight of singleton babies (i.e., without a twin), about 25% of the variation between babies in birth weight is owing to genetic differences. Because of this, and our ability to keep small babies alive, the potency of natural selection acting on this genetic component of the variation has been reduced.

Much Medicine Blunts Selection
Acting on Our Ability to Resist Parasites

While the treatment of newborns has shown great progress, I wonder: What would you nominate as the top medical advance since 1840? This question was posed to medical professionals as well as the public by the

British Medical Journal in 2007. The first edition of the journal was published in 1840, hence their choice of date. What are you going for?

The answers are rather revealing. The top choice surprised me, as it
is something that in the industrialized world is largely taken for granted:
it was the introduction of clean water and sewage disposal.

In the UK the sanitary revolution went hand in hand with a better
understanding of how diseases are transmitted. A turning point was the
investigative work of Dr. John Snow and a local clergyman, Henry
Whitehead, during an 1854 cholera outbreak in London. Prior to their
work, diseases such as cholera and bubonic plague were thought to be
caused by pollution or a form of "bad air"—so-called miasma theory.
Malaria was thought of as being due to a similar cause—hence the name
"malaria," a contraction of Italian *mala aria*, bad air.

From mapping out where cholera victims had lived and where they
obtained their water, Snow and Whitehead discovered a specific water
pump to be the source. Later work discovered that the pump had been
sunk really close to the septic pit of a house. In those days, houses either
had pits for fecal waste or they paid for their waste to be thrown into the
river Thames. The cloth nappy (diaper) of a baby that had contracted
cholera was thrown into the septic pit, and the bacteria (now called
Vibrio cholerae) leaked into the pump's water supply. In that case the
solution was clear (even if the water wasn't): shut down the pump.
The council did this the day after seeing the evidence.

This, however, was only a local fix. It didn't get to the heart of the
problem of recurrent cholera outbreaks. Possibly, then, Snow's greater
contribution came from his analysis of the instances of cholera and the
suppliers of water. As it happens, the way the water supply had developed in London provided a remarkably robust test. Some water suppliers drew their water from the Thames—which, as we have seen, was a
de facto cesspit. Others filtered their water and obtained it from cleaner
sources. As Snow noticed, a house may have a different supplier from
the houses on either side. In addition, the different water companies
that provided the clean and the polluted water supplied the rich and
poor alike, so economic differences could not explain any differences in
cholera rates. As Snow realized, this made for an almost ideal test of
his idea. And in turn, the results were as clear as daylight. The houses

supplied by clean water had an order of magnitude less cholera than those supplied by the dirty water. But it gets even better. To test his theory, Snow turned his attention to prisons and showed that cholera levels went down within days of changing from the dirty water supplier to the clean water supplier. To me this is beautiful, powerful science. Only human brains can do this. Mind you, the proponents of miasma theory were not taking it lying down, and despite the evidence refused to back down. Only human brains do that, too.

In the longer term it was the promotion of proper sewage disposal and clean running water in homes that has saved many lives. But this, the "Great Sanitary Awakening," is not a finished project. In low- and middle-income countries, infantile diarrhea owing to poor sanitation (both removal of human waste and clean water provision) is the second largest cause of death of under-fives, killing 760,000 annually. The number one killer of the under-fives is pneumonia, at 935,000, malaria coming in third.

What would you guess are the next most important medical interventions as voted for in 2007? Coming a close second was the discovery of antibiotics. Anesthesia came in third, vaccines fourth, and the structure of DNA fifth. Next was "germ theory," the idea promoted by Snow in England, Louis Pasteur in France, and Robert Koch in Germany that, in contrast to miasma theory, many diseases were owing to microscopic "germs." It led to a better understanding of the actual causes of disease and thus to their cure and prevention. In the case of listeria-infected milk, for example, the bacteria (*Listeria monocytogenes*) could be killed by heat treatment (pasteurization).

I think this order of breakthroughs, and the killers of children under five in low-income countries, tells you a lot about what kills us without modern medicine. It also tells us about what human genetic variation was most influenced by natural selection, as such selection is mediated much more by early mortality than by what kills you in old age. And for the most part early mortality isn't simply bad genes; it is non-genetic diseases and genetic susceptibility to them. Indeed, if we look for signals in our DNA telling us about past natural selection, these often are seen in genes conferring resistance to parasitic diseases. We have seen direct evidence for such a process in the case of sickle cell anemia. The rapid spread of malaria-resistant forms of the Duffy blood group is another example.

It is unlikely that malaria is the only cause of past selection on our genes. There is also recent evidence for positive selection on the gene *IFITM3* that may well be associated with selection for resistance to influenza. Similarly, in West Africa Lassa hemorrhagic fever, caused by Lassa virus (LASV), is endemic and common. Examination of the blood of Nigerians, looking for antibodies to the virus, suggests that about 20% had been infected. In Sierra Leone and Guinea the number may be more like 50%. And these are the survivors—it is a severe disease and, in some cases, can kill. However, between 50% and 90% of West Africans show few to no symptoms, suggesting that in these populations genetic resistance to the disease may exist. Indeed, if we look at DNA of people from these regions, we see evidence for selection in two of our genes (*LARGE* and *IL21*), both associated with LASV immunity and infectivity. LARGE protein, for example, modifies the cell receptor exploited by the virus to gain entry into our cells. The role of IL-21 isn't so clear, but death from Lassa fever may well involve the sort of cytokine storm made famous by COVID, and *IL21* mutations may make those less likely. We aren't sure.

There are similar suggestions that local infections might cause local selection for resistance. In Bangladesh, for example, many genes showing evidence for selection seem to be involved in resistance to cholera. Indeed, blood type O is very rare in Bangladesh, and the same blood type is associated with a risk of cholera infection becoming severe.

In most of the world this parasite-mediated selection has not gone away. While 17.5% of under-fives die young in low-income countries, only 6% do in industrialized nations. The difference is largely owing to parasites and untreated complications of pregnancy, including downstream effects of maternal mortality, untreated pre-eclampsia, and early infections. The next great set of medical innovations outside of the industrialized nations are then largely implementation of the discoveries that have been applied in the industrialized nations: sanitation, vaccines, antibiotics, and quality care for mothers and babies. Many other interventions are very cheap. Malaria-related mortality has dropped by more than half since 2000 simply by provision of mosquito nets (this would have made my short list of great advances). Oral rehydration therapy for children with diarrhea made it to the top fifteen medical interventions, having saved, it is estimated, 50 million lives over twenty-five years, reducing mortality rates by over 60% per year.

While then the main target of natural selection through our evolutionary history was likely to be resistance to whatever infectious disease was most common and killed us before our time, those in the industrialized world are largely protected from this. People in industrialized nations now die of something else. The World Health Organization estimates that in developed countries, about 30% of deaths are owing to heart disease, 21% to cancer, 14% to stroke, 8% to chronic respiratory diseases, and 7% to violent causes. About 14% of deaths annually are attributed to cigarette smoking, one way or the other.

The more shocking statistic, however, is what causes death in the general population (i.e., not just the under-fives) in low- and middle-income countries. What would you guess? Tuberculosis? Malaria? HIV? Perhaps malnutrition? It's none of these. The WHO estimates that for 9 million deaths in 2012, a remarkable 8.4 million are from pollution. Polluted outdoor air causes about 3.7 million deaths, 4.2 million people died from particulates exposure indoors from cooking stoves, and about 1 million died from chemicals and contaminated soil and water. And 840,000 succumbed to poor sanitation. By contrast, in the same year 625,000 people died from malaria, 1.5 million from HIV/ AIDS, and 930,000 from tuberculosis.

Medicine and Our Impoverished Genes

In this context, it seems decidedly churlish to discuss problems of rare genetic diseases. Indeed, the focus on our genes in this book is largely an irrelevance for most people on the planet. They need clean water more than they need their DNA sequenced. However, genetic diseases and infectious diseases can be coupled. Indeed, some selection for resistance to parasites may well also explain why some genetic diseases are so common. We saw this with sickle cell disease. In addition, seven of eight genes associated with resistance to leprosy, for example, are also associated with risk of inflammatory bowel disease. However, the association is quite complex. One such mutation, in a gene called *NOD2*, increases risk of Crohn's disease but decreases risk of the related disease ulcerative colitis, both inflammatory bowel diseases. Similarly, a variant of the immune gene *ERAP2* both decreases the chance of getting severe pneumonia and increases the propensity to have Crohn's disease. The same variant has also been claimed to have been strongly selected for during the Black Death, but these claims are, to say the least, contentious.

Whatever the merits, in the industrialized West the big money is being directed toward genomics. There are large initiatives to sequence genomes of hundreds of thousands, if not millions, of people in both the UK and the US and then to cross-reference this to their medical history. The premise is simple: we have lots of genetic diseases; if we know which mutations in which genes are causative, then we might also have a hope of finding a therapy.

I have no doubt that useful results will come from such studies—indeed, many already have. But you could be forgiven for being skeptical and thinking that there is more than a whiff of the "Everest" syndrome about such efforts. When asked why people want to climb Mount Everest the answer is often "Because it is there" or "Because we can." In the case of genomics, we certainly can. A complete human genome cost $2.7 billion a few years ago but can now be obtained for under $400 routinely. This will come down even further. Indeed, as of February 2024 the company Ultima Genomics released a sequencing machine that does a human genome for less than $100. You could also be forgiven for being a bit skeptical of the latest "sell" for such endeavors. The case to sequence the first human genome was to know how a human is made and to cure our diseases. It didn't—and couldn't—deliver on either promise. Then, to deliver on these promises, we were told that we needed to understand which genes are expressed and when. Then we needed multiple genomes. Then we needed to understand gene expression in every cell type. I would support such efforts, but I will also concede that the dawn of genomic medicine remains a rather slow dawn: the sun is rising, just not all that fast.

When I survey the last twenty years or so of progress in this field, I am often reminded of the story of the helpful Cub Scout. A person is hunting in grass under a streetlight. The helpful Cub Scout passes by and asks what they are looking for. "My keys," replies the first person. The helpful Cub Scout agrees to help in the search. And so, the two are on hands and knees examining each blade of grass looking for the keys. After a while—and with no success—the Cub Scout asks the obvious question: "Where do you think you lost them?"

"Over there," says the owner of the keys, pointing to a dark patch of long grass.

"So why are we looking here, then?" retorts the Cub Scout.

"Simple," replies the keys' owner. "Because it is light here."

And so it is with much science—we look where we can see, not necessarily where we should be looking to find the keys. The immunologist Peter Medawar put it rather well, describing science as "the art of the solvable." We focus on what we can solve. We can simply, accurately, and cheaply sequence genomes. So we do, despite the fact that the solution to most premature death across the world is not written in our DNA.

But please reserve any skepticism: there is also a reason that DNA came in at number 5 on the list. Our ability to understand infectious disease, and in turn to combat it, is also intimately linked to new sequencing technologies that beat a direct line of descent from the discovery of the structure of DNA. It was, for example, from the sequencing of SARS-CoV-2 in January 2020 that a vaccine could be designed around the sequence of the spike gene within weeks. It is the same DNA technologies that first allowed a glimpse into the causes of cancers and in turn led to drugs to target some of them. In breast cancer, for example, from sequencing, we now think there are at least 13 different genetic "flavors" (for want of a better word). For some of these we have medicines that work, such as Herceptin, which only works for the 30% of breast cancers that overexpress the gene human epidermal growth factor receptor (alias *HER2*). We also know that some mutations (e.g., in genes *BRCA1* and *BRCA2)* are common inherited causes of breast cancer. Sequencing of the DNA of a woman with a family history of breast cancer can then inform her of her risk. This in turn can lead to preventive measures, notably mastectomy (breast removal). This isn't of much use for many, however, as inherited breast cancer is relatively rare (about 5% of all incidences). Most cancer is owing to mutations that occurred during an individual's life.

The idea that genetic diseases come in multiple "flavors" is an important insight, as it highlights that a treatment that works for one may not work for someone else. One of the prospects of sequencing lots of DNA, then, is to enter the world of "personalized medicine" in which the therapy is informed by your individual—often genetic—circumstances. Herceptin is one such example, but more generally, drug therapies tend to be a rather blunt tool. Of particular interest is the avoidance of adverse drug reactions (ADRs). About one in every 16 hospital admissions in the UK is owing to a bad reaction to a medicine, and it leads to occupancy of 4% of hospital beds. Between 10% and 20% of hospital inpatients have an ADR. About 2% of patients admitted with an ADR die. That is a remarkable number—this is a reaction to the drug that was supposed to cure them.

These are at the severe end of effects. For many, a drug has little or no curative effect. The highest-grossing drugs in the USA only benefit between 1 in 4 and 1 in 24 patients. For many there are no obvious effects. For others, the drug makes them poorly, but not bad enough to go to the hospital. There is a reason that your medicines contain a slip of paper telling you about all the side effects you might expect and how common they are.

Part of the variation between people in how they respond to any given drug is often genetic. Wouldn't it be wonderful to be able to say *before* you were given a medicine whether it is likely to work for you, or at the very least, not make you seriously ill? There is hope. Stevens-Johnson syndrome (SJS) and toxic epidermal necrolysis (TEN), for example, are related conditions triggered by certain medicines, such as antiretrovirals, that result in blistering and shedding of the skin. Patients usually end up in burn units. TEN can be so painful many patients want to die. Many do. These conditions are rare but also preventable. And you don't need any fancy technology. There are mutant versions of certain genes that are associated with the condition. The mutant versions of the genes are more prevalent in Asian populations (about 8% of the population of Taiwan, for example, have the predisposing version). Screening for these can be done cheaply, just a few dollars. If you have one of the risk mutations, you should not be given the drug that induces the reaction. In 2011 Dr. Chonlaphat Sukasem at Ramathibodi Hospital in Bangkok came up with a beautifully simple system. Do the screening for this and other drug reactions and enter the tests onto a "pharmacogenetic ID card"—a simple purple wallet-sized card that can be carried about. This can then be shown to your doctor. This has sharply reduced incidences of SJS and TEN in Thailand.

In other cases, the question is not whether to have a drug or not, but rather at what dose. The effects of the blood-thinning drug warfarin, for example, are very sensitive to the levels of the protein CYP450. It was estimated that personalized dosing could prevent 17,000 strokes in the US and eliminate over 40,000 visits to the emergency department. A trial by the Mayo clinic indeed found about a 30% reduction in hospitalizations following personalized dosing.

It's Complex

While I could write about some of the other advances enabled by genetics—and the rapid diagnosis of the metabolic disease that we saw in chapter 2 is an amazing achievement—it is nonetheless a valid question to ask why the sun is only slowly rising on genomic medicine, and in turn to ask what the future might hold. Why are our genetic diseases so hard to conquer?

One reason is that it is complex. I mean this both in an everyday sort of way—cells and organisms are complicated things—and also in a technical way. When it comes to human genetic conditions, we divide them into what are sometimes called Mendelian disorders on the one hand and "complex" disorders on the other. A Mendelian condition is one in which there is a mutation in just one gene that explains whether you have the condition or not. Sickle cell anemia, cystic fibrosis, and

Huntington's disease are like this: one mutation, one gene, one condition. Similarly, for SJS and TEN it is fortunate that knowledge of just one gene variant is a good predictor of an adverse response. For Mendelian conditions, if you track a family history of people, you can see the mutant and normal versions of the genes being transmitted in a simple Mendelian way—half the offspring of an individual with the mutant version and the normal version inherit the mutation.

By contrast, complex conditions involve many genes, each with a mutant version and a normal version. They are also called *polygenic* (literally "many genes") conditions for the same reason. Similarly, much of the variation between people in their height is genetic, but it is owing to small effects of mutations in several hundred, if not thousand, genes. And so it is with many human diseases. Indeed, with the exception of Huntington's disease, neurological conditions, such as schizophrenia, bipolar disorder, or autism, are complex in this sense. Bowel conditions such as ulcerative colitis or Crohn's disease are similarly "complex," as are type 2 diabetes and rheumatoid arthritis. All such conditions may in addition have large non-genetic components. We now know, for example, that multiple sclerosis requires a prior infection with the virus that causes infectious mononucleosis (Epstein-Barr virus). With both environmental and genetic complexity, these are intrinsically harder to understand and to treat.

Before I go further, I should note that the language is a bit silly. The mutations in the genes for complex conditions are also inherited in a Mendelian fashion. We really should just call these simple and complex conditions, not Mendelian and complex. So I will.

For complex conditions, finding the causative mutations and in turn doing something about the conditions is, er, complex. DNA-based technologies have, however, revolutionized our ability to understand such diseases. Since 2007 there has been a major growth in so-called *genome-wide association studies*, GWAS for short. The premise of GWAS is simple. Take loads of people, some with the condition you are interested in and some without it—a so-called case-control method. You then work out which mutations exist in the individuals in the two groups. Mutations that underpin the condition, or that are nearby on the same chromosome as ones that are causative, should be more commonly seen in the individuals with the condition of interest. We thus can get a correlation between a complex disease and particular sets of mutations. The larger the sample size (more people, more mutations), the more powerful the GWAS, hence the desire to sequence hundreds of thousands or millions of genomes (it isn't simply Everest syndrome).

While there have been almost 6,000 GWAS studies of over 3,000 conditions, there is a concern about replicability of GWAS results: If you do the same test using different people, will you highlight the same genes? Some replicable results have been thrown up. The gene *FTO* is repeatably identified in GWAS for obesity, as is *PTPN22* for autoimmune disorders. *FTO* is indeed one of the very few cases where we also have a robust idea of what the gene does, what the mutations then disrupt, and why they cause obesity. There are also concerns, as GWAS is often done on white European populations, given the nature of the sampling. Whether results translate to the rest of humanity is often unresolved. There are attempts to move studies into other groups.

Assuming we have some GWAS hits—mutations that tend to be in individuals with particular diseases—if we wish to probe further, the challenge then is to work out which mutations are actually causative and the functional consequences of these mutations. The downstream causality can be decidedly hard to understand. Sometimes, for example, a mutation in one gene has its effects by changing the transcription of the neighbor gene, not by affecting the host gene. However, we are getting ever better at understanding which mutations might disrupt gene expression (i.e., transcription) and splicing. Work in progress, you might say. For the great majority of GWAS "hits" we have no idea about their mechanism or if they are truly causative.

In a sense this is all phase 1: get to understand the disease. To date this, rather than therapeutics, is where most of the success in GWAS has been found. It is a powerful mechanism to generate further testable ideas as to how a genetic disease comes about. GWAS results, nonetheless, can have considerable utility. Let's look at three of these.

Using GWAS: Drug Repurposing

While successful therapies derived from GWAS are thin on the ground, advocates can point to the utility of GWAS in enabling successful re-purposing of drugs. This is probably where the most important results of GWAS have to date been seen. Drug development is a very expensive and time-consuming business. Drug repurposing takes medicines that pass the safety tests and looks to see if they might be useful for a disease that is different from the one originally intended. This is known as "off-label" medication. This is very attractive to drug companies, as it enables a second market for safe and effective drugs, or can rescue drugs that are safe but don't work for the intended disease. Incidentally, Viagra (sildenafil) was developed to treat high blood pressure and angina but had interesting side effects in many men that led to a redefinition of the market for the drug (it wasn't any good at treating blood pressure).

As GWAS approaches allow us better understanding of underlying mechanisms of disease, they could be important here. There are a few success stories so far. For the inflammatory bowel disease Crohn's disease, GWAS repeatedly identified a key role for the interleukin-23 (IL-23) receptor. This is a protein that sits on cell membranes. When it meets its partner (interleukin-23), it signals to the cells to start an immune response. It looks like Crohn's disease can be due to overactivity of this immune response in the gut. Indeed, one strong GWAS signal was from a mutation in the IL-23 receptor's gene that protects from Crohn's disease. As the mutation seemed to block this signaling, it was thought that a drug to block the signal/receptor interaction could also provide relief from the symptoms.

As it happens, two drugs had been developed and approved that hit this pathway. They are artificially made antibodies to attach to IL-23 and stop it from binding the IL-23 receptor. They had been developed for the skin condition psoriasis. Given the GWAS results, people then thought to see if the same drugs might be good as a Crohn's treatment. Recent randomized controlled trials indeed show that at least one of them is effective. This is now an approved treatment for the disease (if you're interested, the drug goes by the name ustekinumab). It is probably not incidental that Crohn's patients often have psoriasis, as the two conditions share much of the same gene network. Indeed, pleiotropy—where one mutation affects several disorders—is a common feature in GWAS results.

However, by no means need the above history be typical. Indeed, other attempts to transfer GWAS to therapy have not been so successful. GWAS for hypertension, for example, suggested a malfunctioning vitamin D receptor might be to blame. However, in a randomized controlled trial, vitamin D supplementation shows no strong evidence of being useful. For schizophrenia, while we understand the disease much better now, there have been no major changes to drug treatment because of this improved understanding. Incidentally, chlorpromazine (brand name Thorazine), developed in the 1950s for the treatment of schizophrenia, dubbed "psychic penicillin," was nominated as one of the fifteen medical milestones. It has yet to be importantly superseded.

Using GWAS: Risk Scores

Many critics of GWAS, however, suggest that the initial promise of identifying the genes underpinning complex diseases has to some extent been abandoned. Instead, there is a move to a more statistical approach, in which we attempt to define someone's risk of a given condition knowing their variants in toto, so-called *polygenic risk scores* (PRS) or, alternatively, genetic risk scores (GRS). In principle, these approaches don't need the underlying mutations to be causative—they just need to be predictive. The idea, then, is that you could tell people that they are especially at risk of, for example, heart disease. Hopefully that would then stimulate them to a healthier lifestyle. Many genetics companies are betting quite heavily on PRS as being the next big medical intervention.

However, the utility of PRS is contentious. A recent PRS for coronary artery disease was found to miss 85% of individuals with the disease. More generally, a large-scale analysis of 926 genetic risk scores for 310 diseases found that just a little over 10% of individuals who develop the disease in question are identified. At the same time, 5% test "positive," but do not go on to get the disease. By normal medical standards this is pretty poor performance for a test. For something like cardiovascular disease, just going by a patient's age is, it is argued, a much better way to decide whether to prescribe statins. There is understandable skepticism puncturing the hype surrounding PRS.

Nonetheless, in other contexts genetic risk assessment might prove lifesaving. For example, men in their fifties often go to their doctor complaining of problems with their lower urinary tract. They might need to get up in the night to pee, they may need to go very often, or the flow of urine may be weak. Sometimes this is owing to a relatively harmless growth of the prostate gland, at the base of the bladder. The more it grows the harder it is to pee. This is uncomfortable but not life-threatening. However, 3.5% of individuals complaining of these same symptoms will go on to develop prostate cancer within two years. Prostate cancer accounts for about a quarter of all new cancers in men, about 52,000 per year in the UK (and increasing by 4% each year). 10%–20% of deaths could be avoided with earlier detection.

GWAS suggests there to be 269 risk variants for prostate cancer. From these, for any man we can estimate a risk score—the more of the "risky" mutations you have the more at risk you will be. Can we use a genetic risk score using these mutations to decide which of the men should be fast-tracked to the clinic and which can safely not go for further investigation? A recent analysis in the UK suggests it could have some utility. Those with the lowest risk scores (bottom 20%) had less than 1% chance of developing cancer within two years, while those in the top 20% had an 8.8% chance. As an approach, when combined with male age, this is more accurate than the current best test looking at levels of prostate-specific antigen (PSA). It is an open question whether a combination diagnosis using genetics and PSA levels might work even better.

Even if the genetic predictions were used on their own, this could have important consequences for hospital capacity. In the UK, GPs make about 800,000 referrals for possible prostate cancer annually. If we added in information from PRS to this, about 320,000 men could safely avoid referral (which involves both unpleasant investigation and anxiety), while 160,000 could be fast-tracked. This is not yet policy. Indeed, a recent review of all applications of polygenic risk scores found that they may have promise, but clinical benefit has yet to be shown in any case. Watch this space.

Using GWAS: Working Out Causes

Imagine I want to know whether taking some treatment, let's say substance X, lowers your chances of getting some disease, let's say colon cancer. I could do an observational study. Here I would ask whether people who take substance X as a supplement are more or less likely to currently have colon cancer. Imagine that I find that those taking the substance as a supplement are less likely to have colon cancer (substance X is my invention so please don't ask me what this miracle compound might be). Do you think this would this be good evidence that substance X protects you from this cancer?

The possibility that the chemical actually stops you from getting cancer is one possible interpretation of my made-up data. But there are many others. Perhaps some people with the cancer stopped taking the supplement, having gotten ill and believing substance X might be the cause of the cancer. This reverse causality would give the same correlation. Alternatively, people who take substance X may just be more health-conscious and so do all the right things to stay fit and healthy. If so, one of the other things they do—perhaps not eating red meat—could be what explains the low cancer rates in those taking substance X. The study would be confounded by these unknown effects. These problems of causality plague epidemiological observational studies.

The best way to crack the causality problem is a randomized controlled trial (RCT). Here people are chosen at random and told to either take substance X or not. Everything else is random. Some in the substance X group will not eat meat, but so too in the other group. If the substance X group have lower colon cancer rates, then we can draw an inference of causality, as we have broken the confounding issues and the problem of reverse causality. Many such RCTs have indeed overturned results from observational studies, including effects of vitamins and hormone replacement therapy.

While analysis of well-performed RCTs is still the gold standard for inferring causality, it isn't always possible. It can sometimes be impractical or unethical or both. Imagine that you want to know the effects of drinking alcohol over your lifespan. You can hardly select babies at random and instruct them never to drink, while instructing others how much they must drink. In this context a novel approach has developed, based on the back of GWAS studies. Here the idea is simple: if we have a good idea that one or multiple mutations affect the levels of something, then I can allow the rules of Mendelian inheritance to randomize this "treatment" (i.e., mutation) between people. The technique is indeed called *Mendelian randomization* (MR).

An advantage of using mutations is that they break the possibility of reverse causality. Whereas colon cancer might induce people to stop taking substance X, a disease cannot change the DNA you were born with. The randomization should also remove any confounding variables, such as being generally more health-conscious. Unlike asking people to remember how much alcohol they might have drunk in their lifetime, mutations can also be quite easily and accurately assessed. With GWAS data, such genetic data are now routinely available on large, well-characterized groups of people.

MR is now a commonly applied method to try to figure out what works or what causes disease. For example, observational studies suggested levels of C-reactive protein were predictive of coronary heart disease. But are they truly causative? From GWAS a series of mutations were identified that predicted C-reactive protein levels (CRP). However, when comparing people with the different mutations, there was no effect of the mutation (high-CRP vs. low-CRP mutations) on whether individuals subsequently had heart disease. We can then conclude that the prior correlation was most likely owing to some confounding variable. Conversely, the same approach was able to confirm that levels of the low-density lipoprotein version of cholesterol (LDL-C) were indeed not just correlated with, but actually causative of, heart disease.

The technique can also be applied to the effects of behavior. For example, observational studies suggested a link between alcohol intake and raised blood pressure. But there are so many confounding variables here that it is hard to say whether alcohol causes the raised blood pressure or is just correlated with it. Those drinking more could just have generally unhealthy lifestyles, eat too much salt and fat, etc. A full RCT is impractical. However, mutations in the gene *ALDH2* have a major effect on alcohol intake. If you have both copies of the gene with the mutation, alcohol makes you feel sick and so you avoid it. As everything else should be randomized by comparing individuals with high- and low-intake genes, we can obtain a clearer view as to whether it is alcohol intake or something else that causes the high blood pressure association. In this case, the association holds on MR, and so we conclude that alcohol intake is not just correlated with, but causative of, blood pressure changes.

The above cases used single-gene mutations that have a strong association with the key variable (CRP levels, alcohol intake). In principle, multiple strong GWAS-based risk predictors can also be employed, even if we don't have proof that the mutations cause the variation directly—they just need to be good predictors. We can then, in addition, use the fact that there are many mutations in many genes predicting a particular feature to our advantage. For example, nine mutations in six genes we know are associated with different levels of LDL-C. We can score every individual on what you might think of as their genetically determined LDL-C levels based on which versions they have over these six genes. If LDL-C levels truly affect the risk of coronary heart disease, then we can use MR to work out the increase in risk of heart disease for every unit increase in LDL-C exposure over the long term. This indeed confirms a strong causal connection between LDL-C levels and heart disease.

The technique isn't perfect, however. MR can have a problem if a given mutation has effects on other pathways that also have an input to the disease. What if, for example, the mutations in *ALDH2* had direct effects on blood pressure that had nothing to do with avoidance of alcohol? If this were the case, the method could have led us to incorrect conclusions about the relationship between alcohol and blood pressure. We need a good firm coupling between the mutation and the key levels (in this case of alcohol intake) and no alternative pathways. Attempts to remove this "horizontal pleiotropy" problem are ongoing. Using many mutations in many genes may be one way, in the hope that for every gene whose mutations exaggerate the effect there will be another one working in the opposite direction.

The Simple Diseases May Be Easier

While GWAS has exploded as a methodology, in no small part because we have powerful tools, it is not clear that it is yet causing all that great a revolution in healthcare, although we can point to case histories where it has clarified causality, permitted repurposing of drugs, or shown promise for personalized medicine. The sun, as I said, is only slowly rising. As regards drug repurposing, one of the issues is that for many conditions there might not be a single target, because the conditions are complex (in a broader sense). It is striking that GWAS supported the IL-23 pathway, and a patient was also found with something looking like Crohn's disease that presented as a simple (single-gene) disorder. IL-23 receptor may be more than one of many signals—it may be a key pathway. It isn't clear whether other complex diseases might have neat single-gene, or even single-pathway, treatments.

It is also the case that for many complex diseases, not everyone's disease progresses via the same routes. Breast cancer isn't one disease. Similarly, in looking at which genes are expressed in placentas of women with pre-eclampsia, for example, my colleagues and I found that the disease appears to be at least three different conditions, not one.

Given this complexity, perhaps the sun will rise faster for the simple diseases, meaning those where it is one gene, one disease. For these, when we know the gene and know the mutations responsible, we indeed have multiple avenues for therapy, with new ones possibly around the corner. We have multiple ways of defeating the action of natural selection on our mutational burden.

The Royal Disease

To see some options, let's consider a disease that has plagued humans for millennia, the bleeding disorder hemophilia. The tenth-century Arab surgeon al-Zahrawi described families in which boys died of bleeding following just minor traumas. The Talmud instructs that a boy should not be circumcised if he had brothers who died of complications. We now know that this is also not one disease but several. Importantly, each flavor of the disease is associated with mutation in only one gene. The most common form, accounting for 80%–90% of all cases, hemophilia A, affects about 1 in 5,000–10,000 males at birth and is associated with mutations in a gene called *Factor VIII*. Hemophilia B affects one in 25,000 males at birth and is associated with the gene *Factor IX*. There are other flavors, such as hemophilia C, associated with *Factor XI*; von Willebrand disease, associated with the gene for von Willebrand factor; and parahemophilia, associated with *Factor V*. We will consider A and B, as these are the most common.

Note that while there are multiple different genes, this is different from complex diseases where many genes are involved, with different flavors of the disease involving different combinations of multiple genes. With hemophilia, multiple different versions of the disease are each associated with mutations in just one gene, a different gene for each flavor of the disease.

Hemophiliacs have one thing in common: a reduced ability of their blood to clot. When we cut ourselves, the pipework that carries our blood is ruptured. Our clotting mechanism is a delicately balanced system that then springs into action to patch up the cut in the blood vessel. It is a bit like having a system of pipes in your house with the ability to repair from within any leaks (for example, when you hammer a nail by accident into your plumbing). It needs to be balanced because you don't want the blood to clot when this isn't needed—this can lead to blood clots becoming lodged in your lungs, for example. But you do want clotting to start, and start fast, when it is needed. Premature death awaits if the balance either way is wrong.

The way this problem is solved is by having the clotting factors present in a poised but inactive state in your blood, with inhibitors of clotting present, preventing them from triggering a clot. When the blood vessel is cut, this then signals to activate the first protein in the so-called clotting cascade. This first factor then activates the second, which activates the third, etc. Hence the term "cascade." The end product is a fiber-like protein (fibrin) which then plugs the hole. Each of the proteins in the cascade has a name—many are just given a number, like Factor VIII (Factor 8) or Factor IX (Factor 9). If any one of these steps in the cascade is missing, for example owing to mutation affecting the relevant gene, then the cascade doesn't work, and hemophilia is the result.

Moderate to severe hemophilia is most commonly seen in boys, and most commonly of the A or B forms, as the genes for Factors VIII and IX are on the X chromosome. Men have only one X chromosome, so need have only one mutated version of either gene to get the disease. Females with a mutant copy on one X and a normal copy on the other X chromosome are fine (or may have less severe forms).

As females can carry the mutation but be largely fine, the disease can be passed down through families. In both A and B form, brand-new mutations (seen only in the person with the disease) account for about a third of cases. This is testament to our high mutation rate. The other cases inherit the mutation from mothers who were themselves born as carriers. Indeed, hemophilia was once called the "royal disease," as females of nineteenth- and twentieth-century royal families of Europe were often carriers of the mutation, half of their sons thus inheriting the mutation and then getting the disease. Queen Victoria of Great Britain, for example, passed it down to two of her five daughters (Alice and Beatrice), who in turn passed the mutation on to the royal houses of Spain, Germany, and Russia. Victoria's youngest son, Prince Leopold, Duke of Albany, had the disease. He died young, at age 30, following a fall. A simple knock to the head lead to an incurable bleed on the brain. He was lucky. Prior to the 1960s, when the first treatments started to become available, life expectancy was around 11 years. Tests on the bones of the Romanovs killed after World War I (the Russian branch) indicate that they had the rarer B form of the disease, i.e., a mutation in Factor IX. Incidentally, the B form of the disease is also known as Christmas disease, named after the first described patient, five-year-old Stephen Christmas. It was also published in the Christmas edition of the *British Medical Journal.*

What then can we do about hemophilia? Now that we know the causes, one solution is obvious: give the folks without enough Factor VIII protein more Factor VIII protein and give those without enough Factor IX protein more of that protein. This is comparable to the treatment for many metabolic disorders. Indeed, we are all like this—we all lack a functioning *GLO* gene, so we all need vitamin C in our diet. We all suffer the same metabolic disease, scurvy, without dietary input. We just didn't need genetics to discover this and don't need vitamin C injections, just plenty of fruit and vegetables in our diets.

There are several approaches to the supplementation for hemophiliacs. One is to provide blood plasma from folks without the disease (most people). Given the risk of also transmitting infectious agents like viruses, this is often the less desirable option. Indeed, recently for this group the major cause of mortality is no longer a serious bleed, but blood-acquired infections, such as HIV. Another approach is to have injections of either lab-manufactured versions of the proteins or purified forms (i.e., removed from plasma). This is safer, but much more expensive. About a fifth of patients can develop an immune reaction to the supplemented proteins down the road.

As the proteins are also rather unstable, for moderate to severe cases the injections also need to be frequent, up to once a day or once every two days. For less severe cases, the clotting factors can be given when and as needed, i.e., just after a cut or before a tooth extraction, etc. Recently a synthetic mimic of Factor VIII (emicizumab) has appeared on the market. This goes some way to solve the instability issue, the mimics being functionally similar but more stable, requiring injections every few weeks or so.

An alternative approach, now that we know the genetics, is prevention. If a man has the disease, his sons—who inherit his Y chromosome, not his X chromosome—should be fine. A carrier female's daughters should also be fine (they could be carriers), but half of her sons would not be fine.

But what can we do with this information? There are many possibilities. One approach is to opt for someone else's genes, either via sperm or egg donation or via adoption. Another is the possibility of prenatal diagnosis. This can either be via chorionic villus sampling (CVS), typically done during weeks 10–13 of pregnancy, or amniocentesis, at about 15 weeks. CVS takes a small section of the placenta for testing, while amniocentesis takes the liquid—which has fetal cells floating in it—that surrounds the baby. Both have small risks (1% miscarriage risk for CVS, 0.3% for amniocentesis). What a family will do with such information is then up to discussion (legal restrictions aside), but could include selective abortion or preparatory work: i.e., if you know your son will have hemophilia, you can be prepared—both physically and mentally—for what is to come.

Another option is so-called *pre-implantation genetic diagnosis* (PGD). This is made possible by now-commonplace in vitro fertilization (IVF). The mother's eggs are removed and fertilized in the lab using the father's sperm. We then have a panel of embryos, minuscule bundles of cells. The possibility then is to work out, by extracting one cell from each embryo, whether the embryo has the mutation or whether it is male (meaning XY) or female (XX), and then to implant back into the mother only the chosen embryos. This, like selective abortion, is unnatural—but potentially beneficial—selection, you might say.

Naturally, these technologies bring with them numerous ethical issues. PGD requires the selective removal of some embryos. With recent advances in technology, we also now have the ability to take the rejected embryos, those with hemophilia, and from these derive stem cell lines that can be used for the study of hemophilia. Does this alter your moral judgment? Does your calculation of the ethics take into account the alternative treatment options for any son born with hemophilia? What if we are considering an X-linked disease that kills early and for which there is no "cure"?

Would you approve PGD if only the next generation was to be prevented from having hemophilia? I know of one case where a family with a father that was affected and a mother who was not a carrier asked for PGD, even though the daughters would be carriers and the sons unaffected. They wanted to rid their lineage of the disease and so sought to implant only the sons. Would you have approved this had you been a medical regulator? The request was declined, as the process was more one of manipulation of mutational frequencies (i.e., eugenics) rather than the prevention of immediate harm. But there could be harm in the second generation when the carrier daughters have sons. Would you consider this to be relevant?

Would it make any difference to you if the same end was achieved by sorting out an affected male's sperm into X- and Y-bearing ones and only using the Y bearing ones? Or, if the female is a carrier, by only using an unaffected father's X-bearing sperm? This technology is now possible, and could be aligned with a simple procedure in which a mother ovulates naturally and the selected sperm are injected into the uterus. This is most assuredly messing with genetics, but not at the cost of any embryos. Is that ethically different? Would you approve of this?

In the case of "simple" genetic diseases (i.e., one mutation in one gene gives the disease), these issues are hard enough. We can now also do something similar for complex diseases. We can employ polygenic risk scores to select from a panel of IVF embryos the ones least likely to have the complex disease in question. However, as the panel of embryos is limited, the technique is not especially effective. In the UK it is, as a consequence, illegal. In the US, by contrast, it is legal; firms have grown up to offer such services. The first baby selected after IVF-mediated polygenic risk score screening was born in 2020.

Gene Therapy—the Next-Generation Fightback against Mutation

Into these ethical minefields comes a new option. All the above attempt to minimize the impact of the disease by providing the missing protein or to prevent the disease by artificial selection. But what if you could do something else—repair the gene or otherwise give the patient a functioning version of the gene? This is the brave new world of gene therapy. It is exciting and dangerous at the same time.

In the case of hemophilia, the premise behind gene therapy is that we could replace the very common transfusions, needed for all the life of a sufferer, with one treatment. This treatment gives them the gene and with it the ability to generate the protein product. It sounds simple. In practice it is anything but, with both technological and safety hurdles to be overcome. These start with what is called the delivery problem: How are you going to get a gene synthesized in the lab into a patient, in the right cells? After that, we need to make sure it is expressed at the right time and doesn't do any harm.

The field has had what might be described as a long and bumpy road. That, however, doesn't do justice to folks who died along the way. A cornerstone analysis from Theodore Friedmann and Richard Roblin in 1972 saw all the potential for incorporating DNA as a treatment. They, however, strongly urged caution and explicitly opposed any gene therapy, at least for the foreseeable future. They had three problems. First, we didn't at the time understand how DNA in our cells might get swapped about or how genes are turned on or off. Second, we didn't really understand well how given mutations led to disease states. Third, we had no idea about side effects, either in the short or the longer term.

Eighteen years later the first successful gene therapy was reported. A four-year-old girl, Ashanthi DeSilva, lacked a key enzyme, adenosine deaminase (ADA), without which her immune system was all but non-functional. Almost any infection could kill her. Her treatment involved taking a functioning version of the gene, putting it into a modified version of a virus, and giving her cells the modified virus. Viral "vectors" like this are often used, as viruses have evolved the capability of getting into our cells. This was a success story. DeSilva survived to live a normal life.

Enthusiasm was, as you might predict, high. However, this was premature. Nine years later (1999), gene therapy sent to cure killed its first patient during a clinical trial. In this case, an eighteen-year-old with a mutation that meant his liver couldn't break down ammonia—which then accumulated in his blood—died only a few days after being given an adenovirus vector (a bit like a cold virus) with a good version of the gene. He suffered a massive and fatal immune reaction to the viral vector. In 2006, three of twenty patients given a gene therapy for X-linked severe combined immunodeficiency died. In this case the gene therapy involved DNA to be introduced into the patients in cells that make up the immune system of the blood. In some cases, however, the gene inserted into the patients' DNA in such a manner that it disturbed the expression of another gene (called *LMO2*), and this, it seems, caused the patients to develop cancer. On discovering this, the instigators of the clinical trial asked the French authorities to halt their trial. They just didn't understand the risks.

These setbacks stimulated a large amount of soul-searching, as they should. Had gene therapy gone too fast? Had the complications not been understood, as Friedmann and Roblin had warned?

Despite these setbacks, the field is back. At the last count there are nearly 3,000 gene therapy trials underway, and the US FDA expects to be approving 10 to 20 such therapies by 2025. One of the key problems, the fact that the vector itself triggers an immune response, seems to have been reduced to a large degree. The new superstar vectors are so-called *adeno-associated viruses* (AAVs). They seem fairly safe and can give long-term expression of the desired gene. However, almost as soon as I wrote these words there came a report of two children dying of acute liver failure after receiving gene therapy for spinal muscular atrophy, Zolgensma, that used an AAV vector targeting the liver. Liver failure is a known side effect, reported in about a third of patients.

People with heritable blindness have been some of the first beneficiaries. The eye is a closed system and so intrinsically safer than many other tissues to work on. You can inject the virus with the gene into the back of the eye and it will stay in the eye. This makes the therapy much safer and is thus attractive to drug developers. In 2018 Luxterna was approved for use in Europe to treat heritable blindness owing to mutations in a gene, *RPE65*. The mutation causes the light-receptive cells to die over time. Giving the same cells the "correct" version of the gene showed improvements in vision as early as 30 days after injection. At the time, at $850,000 per treatment it was the world's most expensive drug.

The eye-watering costs (if you forgive the dreadful pun) raise a further hurdle for gene therapy. In 2012, a gene therapy for the rare disease lipoprotein lipase deficiency was approved in the EU. However, as it cost a million euros and was prescribed for only one patient, the treatment was withdrawn in 2017. It simply wasn't economically viable for the manufacturer. Even if in principle gene therapy might be applied to any of the multiple rare simple diseases, it is unclear whether the numbers will ever add up, just because each one is rare. You can imagine that more common disorders like sickle cell anemia and cystic fibrosis are economically more viable targets.

Perhaps you will not be surprised to hear that gene therapy for both hemophilia A and hemophilia B are now in process. In 2022, the European Medical Agency granted conditional market approval for Roctavian, a gene therapy for hemophilia A. It doesn't apply to all patients. This is an AAV vector, and the patients must not have antibodies to the viral vector. This should avoid the worst of the powerful immune reactions to the therapy. Some patients have also developed an immune response to Factor VIII. It can't be used for them either. For those who could have the treatment, the idea is that the vector transports the gene to the liver and there enables the patient to make their own Factor VIII. Two years after the therapy the bleeding rates were reduced by 85%, and 128 of 134 male patients no longer need the replacement therapy. There can be liver damage as a side effect, but this can be treated with corticosteroids.

For hemophilia B similar progress is also being reported. In late 2022, the US FDA approved Hemgenix (alias etranacogene dezaparvovec) as gene therapy for hemophilia B (at the modest price of $3.5 million per treatment). This provides a copy of the Factor IX gene bundled into an adeno-associated virus vector that goes to the liver. In a study with 54 patients, Factor IX levels went to just under 40% of the average "normal" level, which itself can vary from 50% to 150% normal. Forty percent would be considered mild hemophilia. Indeed, the treatment resulted in decreased need for the replacement infusions and a more than 50% drop in bleed events. In another trial, nearly all patients (94%) no longer needed the infusions at all.

It isn't, however, all good news. There were side effects such as flu-like symptoms, tiredness, and generally feeling unwell. It also didn't work for all. Some subjects showed only modest rises (5%–10% of normal levels), while some even exceed the normal average level. The causes of this variation are a mystery. As with the Factor VIII therapy, it also cannot be given to all patients, notably those with inhibitors of Factor IX.

Hemgenix isn't the only gene therapy for Factor IX. LT180a is also a liver-directed AAV with Factor IX as the cargo. Early trials reported in 2022 suggest it is well tolerated, although patients are given immuno-suppressants to counter any immediate immune reaction. Nine of ten patients showed increased Factor IX levels after half a year, but all patients had some sort of adverse effect.

And what of sickle cell disease? As it is a condition in which severe disease is associated with having two copies of the mutant version of the gene, this is a strong candidate for successful gene therapy. Here a slightly different approach is being employed. If you recall, the disease is owing to a single mutation in beta globin. As it happens, as fetuses we don't express beta globin, but a related gene, gamma globin. As we develop, this is then inactivated by a gene called *BCL11A* and we express beta globin instead. What if we could turn on the fetal gamma globin gene by reducing BCL11A levels or by stopping BCL11A-mediated repression? Would this lead to high levels of gamma globin, and would this be able to substitute for the mutant beta globin? Using CRISPR-based gene editing technologies (that we met in chapter 6), researchers have managed to delete part of the DNA that functions as the "on" switch for *BCL11A* in blood cell precursors. The individuals treated this way did indeed then make more gamma globin. This recently underwent clinical trials and the treatment, Casgevy, is as of late 2023 an approved treatment in the UK and US (it was approved in the EU in early 2024). This means it is deemed effective and safe, but not necessarily cost-effective. Indeed, in the UK, as of spring 2024 it was not recommended for funding to treat sickle cell disease given the cost (in the US it is more than $2 million per treatment). As of August 2024, it was, however, recommended for funding in the UK to a limited degree for the treatment of a different condition, transfusion-dependent beta thalassemia, this also being a genetic defect in beta globin, but one—as the name suggests—requiring regular (every 3–5 weeks) blood transfusions. In early trials more than 90% of beta thalassemia patients required no transfusions for at least a year after the therapy.

Lyfgenia is a competitor gene therapy, FDA-approved also in December 2023. This is a more "classical" gene therapy: a viral vector that is modified to incorporate an anti-sickling version of beta globin. This treatment, however, comes with a "black box" warning on the label to draw attention to fact that a few of the treated patients developed blood cancer.

Here Be Dragons:
The Perils of Germline Gene Therapy

There seems little doubt that gene therapy in one form or other is the future for some single-gene disorders. Notice, however, that the therapy is to cure the patient alone. This is known as *somatic therapy* and has no effects on the kids of the patient. Much more contentious is germline therapy. This one could do by correcting the gene in IVF embryos. While in somatic therapy the patient alone is the one affected, the individuals that develop from IVF-modified embryos in turn can transmit the corrected version to their offspring. Germline therapy is really messing with human genetics.

Despite profound ethical issues, it has, however, been tried. In November 2018 researcher He Jiankui announced by YouTube video just before the International Summit on Human Genome Editing at which he was speaking, that he had, using CRISPR technology, edited many embryos fathered by men affected with HIV, and reported the birth of twins (given the pseudonyms Lulu and Nana) with an altered version of the gene *CCR5*. A certain mutant version of this gene (called *CCR5-delta 32*) blocks HIV, and the aim of the editing was to give the edited embryos this version. Note that this was gene editing—changing the resident gene—rather than introduction of a brand new gene, but both can sit under the gene therapy umbrella.

The response to this announcement rather well captures the Wild West that gene therapy can easily become. The first announcements from the Chinese media proudly announced a great breakthrough. You will struggle to find these headlines by Google searching, as they were rather shortly after taken down. There was within the genetic community a profound uproar. Beyond the level of secrecy that had surrounded

He's work, the problems were multiple. Most would argue that HIV is presently not a serious enough condition to be considered for embryonic genetic manipulation. There are guidelines for contemplating such embryonic gene modification, foremost of which is that it must meet a "serious, unmet medical need." A few diseases currently might fit that classification: Huntington's, cystic fibrosis, Tay-Sachs disease being three, familial Alzheimer's a possible fourth. HIV infection does not, as there are good therapies. He disagreed and, according to the ethics form submitted to his hospital, considered his work "more significant than IVF."

However, just about everyone in the field agrees that the technology just isn't there yet, even if the ethical case is otherwise sound. CRISPR has a tendency to affect not only the gene targeted but other parts of the DNA as well, causing hard-to-predict "off-target" effects. How can you make sure that you "do no harm"? Furthermore, if you are giving the CRISPR chemical cocktail to multiple cells, not all cells are affected by it, and not all in the same way. While He's manipulations were done at the one-cell stage, there is evidence that some effects carried over to the multiple-cell stage. It is then possible that an embryo ends up a mosaic of cells and the offspring not immune to HIV. Even if the edit had no off-target effects, there are also indications that having mutant $CCR5$ has effects beyond blocking HIV. Mutant $CCR5$ also affects brain function— mice with the gene knocked out have improved memory, and people with a mutant version recover faster from strokes. Ironically, after He's conference presentation, the next talk at the Congress was about what the pathway to germline gene therapy/editing should look like, why we aren't there yet, and what the hurdles are.

While He featured in the pioneer list of *Time's* 100 most influential people, rather than being a candidate for the Nobel prize (as he suggested on his ethics form), he was fired and landed in prison for three years. What do we make of this? Many think He is a rogue scientist driven by personal ambition and that his actions were irresponsible, premature, and dangerous. One hundred twenty-two Chinese scientists condemned his work as "crazy" and considered it to be a serious dent both in the development of science in China and in its global reputation.

Others, however, think him a sacrificial lamb. Germline therapy is currently banned in more than 70 countries. But rumors persist that it is being tried.

Safe Harbors: Where Evolution and Gene Therapy Meet

Even with the somatic therapies, we must wonder if we are still playing with fire. For some gene therapies the intent is for the new DNA to be inserted into the DNA of the patient. This isn't the case with Hemgenix, but alternative possible gene therapies put the gene in question into a vector that both delivers the gene to the cell and integrates it into the patient's DNA. The rather wonderfully titled *Sleeping Beauty* is one such delivery system, and is being actively investigated as a means to treat hemophilia. This must come with many worries. Echoing those earliest concerns, do we really understand genomes well enough to know what is safe? That the insertion of the therapeutic gene for X-linked severe combined immunodeficiency is what killed some of the patients came as quite a wake-up call. The original gene therapy attempts made some bold assumptions: that 90% of the DNA was probably functionless, that integration sites would be random, and that the activity of one gene was unlikely to affect that of others. In combination, it was assumed that gene insertion, should it take place, would not be problematic.

All assumptions can be questioned. We now know that if you insert a new gene, it often will itself be affected by the neighboring genes and also potentially affect the neighbors. Gene therapy is here in a bit of a bind: you need the new inserted gene to be well expressed (transcribed), but you don't want this expression to affect the neighbor genes. Given such issues, there is discussion of so-called *genomic safe harbors*. These are hypothetical locations in our DNA where a gene could insert and not affect neighbor genes, not cause cancer or other unwanted effects. One of the reasons *Sleeping Beauty* is attractive as a choice of transport is that it tends not to insert next to genes, while many other viral transport packages often insert very close to active genes. That such incorporation happens even if vectors are designed not to incorporate at all only questions the safety of all modes of gene therapy (insertional and non-insertional).

In cases where cells are extracted for the addition of the gene in the lab prior to reintroduction—as in some gene therapies involving blood cells—one could imagine screening cells to see which might be safe. At least it is now usual to make sure no cell divides much better than any other, as such better cell lineages could easily give rise to a cancer. In cases—as in the hemophilia treatments—where we inject the virus and it goes to the liver, we can think of asking which vectors are likely to be safer than others, more likely to target safe harbors. We can also try to make the vectors safer by building in bits of DNA that prevent effects on the genomic neighbor genes.

But what might a safe harbor location look like? There are attempts to define what a safe insertion site position might be like—it cannot be nearer than a certain distance from the nearest gene, for example. The definition tries to identify what we assume to be functionally irrelevant DNA. But what if we are wrong and we simply haven't understood the function of all that seemingly superfluous DNA? The definition of a safe harbor location has indeed been changed since its first outing, as it was clear we didn't really understand. Even as recently as 2020, a new gene added using a so-called lentiviral vector (the one thought less likely to be disruptive) caused cancer in just under 10% of experimental mice. As of February 2021, a trial of gene therapy for sickle cell anemia was paused after two patients developed a serious illness, acute myeloid leukemia, while on the trial. Whether the therapy was causative is not clear.

But, it must be asked, do we have a good enough understanding to know which regions of our DNA would be safe to insert a new gene into? Some introns may indeed regulate other genes when spliced out. Would inserting a new gene into one such intron be harmful? Perhaps some of the seemingly functionless DNA is needed, perhaps for structural reasons? Many parts of the genome make non-coding transcripts that are fast-evolving, suggesting that the sequence is not under selection. But perhaps the process of transcription, rather than the product, is under selection? Could this explain why so many transcripts are made but then are sent directly to the recycling bin?

One can always invent adaptive hypotheses for anything—for why it is there. But if it were not there, we could come up with adaptive hypotheses for that, too. Hypotheses are not themselves good enough and just measure our ingenuity. The bigger issue is whether there is any reason to suppose we might be missing something.

And this brings us full circle to the problem of how our genome evolves. The data seem to support, as I have argued, the notion that bloated genomes are full of junk that we cannot get rid of because selection is an inefficient process when population sizes are small. There is, however, one big unsolved issue here: What explains organismic complexity? For all the stupidity of our genes and genome, something must explain why we have so many cell types, our best measure of complexity. Is it in the non-coding RNAs? Or is it perhaps in the reels of film that get spliced but never get involved in the making of protein? Is it in the diversity of splice forms? We can, after all, be confident that not *all* are junk.

Small population size and weakened selection can explain why our introns are longer than those of mice, for example, but broadly viewed, we still don't know what explains complexity. And there is a problem trying to find out: species with small populations, like us and elephants, tend to be more complex than ones with large population sizes, like flies, bacteria, and yeast. As such, unpicking the causes of complexity from the consequences of chance is tricky. Is our genome so complex because we are complicated beings, or is it a necessary consequence of our small population size, weakening selection's efficacy? How much of our DNA is functional remains a thorny issue, not helped by biased gene conversion, which messes up lots of methods to try to work it out. Its activity leaves footprints in the DNA—like slow evolution—that look a lot like selection's activity.

Should then we be risking gene insertion without a better understanding of how our genome works? One solution is to do gene therapy but avoid inserting a new gene—just get the DNA into cells. Even here, there is a worry that a gene that isn't intended to be inserted into DNA might do so anyway. AAVs are for the most part not intended to insert into DNA, but in animals evidence suggests that they can. I am concerned. Put differently, I have looked at genomes for many years now. I am quietly confident that we have the big picture. Our genome is large and bloated largely because selection is weak and inefficient when population sizes are small. But if you showed me a position where a gene had inserted into my DNA and asked me whether I thought it was safe, I would be anxious. I don't think I could be confident in my judgment, safe harbors or not. Proceed with caution would be my advice.

Afterword

Medicine, then, especially the new age of genetic medicine, has the prospect of holding back natural selection either by keeping us alive despite our burden of mutations, or by artificial selection favoring embryos without diseases. The weakened force of natural selection when population sizes are small can be tamed—at least to some degree—by that other product of evolution: our amazing brains.

And with these same brains, what in the round are we to make of this new view of what it is to be human? Our DNA is largely junk; we have a high mutation rate and many problems making babies. How does that change your view of what it is to be human? That great march of progress, as we saw in chapter 1, looks like wishful thinking more than objective reality. But does that make human existence less valuable or more valuable? How does it alter, if at all, your view of what it is to be human?

For me, it enables a humbler view of what it is to be *Homo sapiens*. That march of progress captures so many of our presumptions about our intrinsic assumed greatness that anything to disturb that view is probably for the good. Are we, as a species, the best nature has to offer, the top of the tree? Not at all. Perhaps the humility that comes with this view might change your view of the rest of life, the species that we share the planet with.

In addition, the fact that we are not the embodiment of perfection makes the fact that we are alive all the more remarkable. There are many, many humans that never made it from egg to baby. Many others die an early death. We are the lucky few alive to read this. We are the survivors, despite our impoverished, mutation-ridden genomes.

In addition to this, there is a feature of many organisms, humans included, that I never cease to be amazed by. It is the strangest of strange facts about us: we are somehow born young. That might sound a bit odd. By definition babies are born at age zero. What I mean is different. When we make babies, we combine the DNA of two old people. Indeed, the eggs that mothers provide are as old as the mother herself—they are held in a static state from birth to adulthood. How can two old people combine together to make a baby that is born young not simply in years but also in life potential? By life potential I mean the number of years it might expect to be alive. For example, I am 60 now, but if I had a baby it should live for 70 years or more while I have only a few years left. We are all born with the same three-score-years-and-ten potential (give or take). The slate of life has somehow been wiped clean, and each new baby starts out afresh.

It doesn't have to be like this. For example, if you keep our cells growing in the laboratory, they just get worse and worse over time—they can only divide 40–60 times before they cannot divide anymore and break down and die. But this ability to be refreshed through sexual reproduction is true in us and in single-celled sexual species—somehow meiosis allows organisms to wipe the slate clean. The development of all organisms is rather wondrous like this. It still seems to me one of the greatest of all marvels the way a single fertilized egg develops into a new, fresh life. It is worth understanding that while we are not the pinnacle of evolution, nonetheless the marvel of life remains glorious, despite our imperfections.

BIBLIOGRAPHY

Further Reading for Chapter 1

Baggini, J. 2004. *Making Sense: Philosophy behind the Headlines.* Oxford University Press.

Cohen, A. 2016. *Imbeciles: The Supreme Court, American Eugenics, and the Sterilization of Carrie Buck.* Penguin Books.

Cook, L. M., Mani, G. S., and Varley, M. E. 1986. "Post-industrial melanism in the peppered moth." *Science* 231:611–613.

Daly, M., and Wilson, M. 1988. "Evolutionary social psychology and family homicide." *Science* 242:519–524.

Dawson, G. 2024. *Monkey to Man: The Evolution of the March of Progress Image.* Yale University Press.

Maynard Smith, J. 1978. *The Evolution of Sex.* Cambridge University Press.

Okasha, S. 2020. "Biological Altruism." *The Stanford Encyclopedia of Philosophy,* Zalta, E. N., ed. https://plato.stanford.edu/archives/sum2020/entries/altruism-biological.

Further Reading for Chapter 2

Bach, J.-F. 2018. "The hygiene hypothesis in autoimmunity: The role of pathogens and commensals." *Nature Reviews Immunology* 18:105–120.

Belinky, F., Babenko, V. N., Rogozin, I. B., and Koonin, E. V. 2018. "Purifying and positive selection in the evolution of stop codons." *Scientific Reports* 8:9260.

Fan, S., Hansen, M.E.B., Lo, Y., and Tishkoff, S. A. 2016. "Going global by adapting local: A review of recent human adaptation." *Science* 354:54–59.

Owen, M. J., Niemi, A.-K., Dimmock, D. P., Speziale, M., Nespeca, M., Chau, K. K., et al. 2021. "Rapid sequencing-based diagnosis of thiamine metabolism dysfunction syndrome." *New England Journal of Medicine* 384:2159–2161.

Plomp, K. A., Viðarsdóttir, U. S., Weston, D. A., Dobney, K., and Collard, M. 2015. "The ancestral shape hypothesis: An evolutionary explanation for the occurrence of intervertebral disc herniation in humans." *BMC Evolutionary Biology* 15:68. https://doi.org/10.1186/s12862-015-0336-y.

Rice, W. R. 2018. "The high abortion cost of human reproduction." *bioRxiv* 372193. https://doi.org/10.1101/372193.

Further Reading for Chapter 3

Baym, M., Lieberman, T. D., Kelsic, E. D., Chait, R., Gross, R., Yelin, I., and Kishony, R. 2016. "Spatiotemporal microbial evolution on antibiotic landscapes." *Science* 353:1147–1151.

Haig, D. 1993. "Genetic conflicts in human pregnancy." *Quarterly Review of Biology* 68:495–532.

Lenski, R. E. 2017. "Experimental evolution and the dynamics of adaptation and genome evolution in microbial populations." *ISME Journal* 11:2181–2194.

Manceau, M., Domingues, V. S., Mallarino, R., and Hoekstra, H. E. 2011. "The developmental role of Agouti in color pattern evolution." *Science* 331:1062–1065.

Trivers, R. L. 1974. "Parent-offspring conflict." *American Zoologist* 14:249–264.

Further Reading for Chapter 4

Kimura, M. 1983. *The Neutral Theory of Molecular Evolution.* Cambridge University Press.

Ohta, T., and Gillespie, J. H. 1996. "Development of neutral and nearly neutral theories." *Theoretical Population Biology* 49:128–142.

Further Reading for Chapter 5

ENCODE Project Consortium. 2012. "An integrated encyclopedia of DNA elements in the human genome." *Nature* 489:57–74.

Graur, D. 2017. "An upper limit on the functional fraction of the human genome." *Genome Biology and Evolution* 9:1880–1885.

Graur, D., Zheng, Y., Price, N., Azevedo, R. B., Zufall, R. A., and Elhaik, E. 2013. "On the immortality of television sets: 'Function' in the human genome according to the evolution-free gospel of ENCODE." *Genome Biology and Evolution* 5:578–590.

Lynch, M. 2006. "The origins of eukaryotic gene structure." *Molecular Biology and Evolution* 23:450–468.

Lynch, M. 2007. *The Origins of Genome Architecture.* Sinauer Associates, Inc.

Ponting. C. P., and Haerty, W. 2022. "Genome-wide analysis of human long noncoding RNAs: A provocative review." *Annual Review of Genomics and Human Genetics* 23:153–172.

Rands, C. M., Meader, S., Ponting, C. P., and Lunter, G. 2014. "8.2% of the human genome is constrained: Variation in rates of turnover across functional element classes in the human lineage." *PLoS Genetics* 10:e1004525.

Zhang, J., and Xu, C. 2022. "Gene product diversity: Adaptive or not?" *Trends in Genetics* 38:1112–1122.

Further Reading for Chapter 6

Allison, A. C. 2009. "Genetic control of resistance to human malaria." *Current Opinion in Immunology* 21:499–505.

Cagan, A., Baez-Ortega, A., Brzozowska, N., Abascal, F., Coorens, T.H.H., Sanders, M. A., et al. 2022. "Somatic mutation rates scale with lifespan across mammals." *Nature* 604:517–524.

Codoñer, F. M., Darós, J.-A., Solé, R. V., and Elena, S. F. 2006. "The fittest versus the flattest: Experimental confirmation of the quasispecies effect with subviral pathogens." *PLOS Pathogens* 2:e136.

Galipeau, P. C., Oman, K. M., Paulson, T. G., Sanchez, C. A., Zhang, Q., Marty, J. A., et al. 2018. "NSAID use and somatic exomic mutations in Barrett's esophagus." *Genome Medicine* 10:17. https://doi.org/10.1186/s13073-018-0520-y.

Horton, J. S., Flanagan., L. M., Jackson, R. W., Priest, N. K., and Taylor, T. B. 2021. "A mutational hotspot that determines highly repeatable evolution can be built and broken by silent genetic changes." *Nature Communications* 12:6092.

Lynch, M., Ackerman, M. S., Gout, J.-F., Long, H., Sung, W., Thomas, W. K., and Foster, P. L. 2016. "Genetic drift, selection and the evolution of the mutation rate." *Nature Reviews Genetics* 17:704–714.

Melamed, D., Nov, Y., Malik, A., Yakass, M. B., Bolotin, E., Shemer, R., et al. 2022. "De novo mutation rates at the single-mutation resolution in a human *HBB* gene region associated with adaptation and genetic disease." *Genome Research* 32:488–498.

Pal, C., Maciá, M. D., Oliver, A., Schachar, I., and Buckling, A. 2007. "Coevolution with viruses drives the evolution of bacterial mutation rates." *Nature* 450:1079–1081.

Sprouffske, K., Aguilar-Rodríguez, J., Sniegowski, P., and Wagner, A. 2018. "High mutation rates limit evolutionary adaptation in *Escherichia coli.*" *PLoS Genetics* 14:e1007324.

Wang, L., Sun, Y., Sun, X., Yu, L., Xue, L., He, Z., et al. 2020. "Repeat-induced point mutation in *Neurospora crassa* causes the highest known mutation rate and mutational burden of any cellular life." *Genome Biology* 21:142. https://doi.org/10.1186/s13059-020-02060-w.

Wang, L., Ji, Y., Hu, Y., Hu, H., Jia, X., Jiang, M., et al. 2019. "The architecture of intra-organism mutation rate variation in plants." *PLoS Biology* 17:e3000191.

Wang, Y., and Obbard, D. J. 2023. "Experimental estimates of germline mutation rate in eukaryotes: A phylogenetic meta-analysis." *Evolution Letters* 7:216–226.

Further Reading for Chapter 7

Akera, T., Trimm, E., and Lampson, M. A. 2019. "Molecular strategies of meiotic cheating by selfish centromeres." *Cell* 178: 1132–1144.e10.

Amorim, C.E.G., Gao, Z., Baker, Z., Diesel, J. F., Simons, Y. B., Haque, I. S., et al. 2017. "The population genetics of human disease: The case of recessive, lethal mutations." *PLoS Genetics* 13:e1006915.

Cocquet, J., Ellis, P.J.I., Mahadevaiah, S. K., Affara, N. A., Vaiman, D., and Burgoyne, P. S. 2012. "A genetic basis for a postmeiotic X versus Y chromosome intragenomic conflict in the mouse." *PLoS Genetics* 8:e1002900.

Haeusler, M., Grunstra, N.D.S., Martin, R. D., Krenn, V. A., Fornai, C., and Webb, N. M. 2021. "The obstetrical dilemma hypothesis: There's life in the old dog yet." *Biological Reviews* 96:2031–2057.

Haig, D. 2000. "The kinship theory of genomic imprinting." *Annual Review of Ecology, Evolution, and Systematics* 31:9–32.

Hastings, I. M. 2001. "Reproductive compensation and human genetic disease." *Genetical Research* 77:277–283.

Hurst, L. D. 2022. "Selfish centromeres and the wastefulness of human reproduction." *PLoS Biology* 20:e3001671. https://doi.org/10.1371/journal.pbio.3001671.

Moore, T., and Haig, D. 1991. "Genomic imprinting in mammalian development: A parental tug-of-war." *Trends in Genetics* 7:45–49.

O'Brien, E. K., and Wolf, J. B. 2017. "The coadaptation theory for genomic imprinting." *Evolution Letters* 1:49–59.

Further Reading for Chapter 8

Cáceres, E. F., and Hurst, L. D. 2013. "The evolution, impact and properties of exonic splice enhancers." *Genome Biology* 14:R143. https://doi.org/10.1186/gb-2013-14-12-r143.

Duret, L., and Galtier, N. 2009. "Biased gene conversion and the evolution of mammalian genomic landscapes." *Annual Review of Genomics and Human Genetics* 10:285–311. https://doi.org/10.1146/annurev-genom-082908-150001.

Hartfield, M., and Keightley, P. D. 2012. "Current hypotheses for the evolution of sex and recombination." *Integrative Zoology* 7:192–209. https://doi.org/10.1111/j.1749-4877.2012.00284.x.

Ho, A. T., and Hurst, L. D. 2022. "Unusual mammalian usage of TGA stop codons reveals that sequence conservation need not imply purifying selection." *PLoS Biology* 20:e3001588. https://doi.org/10.1371/journal.pbio.3001588.

Lively, C. M. 1987. "Evidence from a New Zealand snail for the maintenance of sex by parasitism." *Nature* 328:519–521. https://doi.org/10.1038/328519a0.

Morran, L. T., Schmidt, O. G., Gelarden, I. A., Parrish, R. C., II, and Lively, C. M. 2011. "Running with the red queen: Host-parasite coevolution selects for biparental sex." *Science* 333: 216–218. https://doi.org/10.1126/science.1206360.

Further Reading for Chapter 9

Abdellaoui, A., Yengo, L., Verweij, K.J.H., and Visscher, P. M. 2023. "15 years of GWAS discovery: Realizing the promise." *American Journal of Human Genetics* 110:179–194.

Bulaklak, K., and Gersbach, C. A. 2020. "The once and future gene therapy." *Nature Communications* 11:5820. https://doi.org/10.1038/s41467-020-19505-2.

Green, H. D., Merriel, S.W.D., Oram, R. A., Ruth, K. S., Tyrrell, J., Jones, S. E., et al. 2022. "Applying a genetic risk score for prostate cancer to men with lower urinary tract symptoms in primary care to predict prostate cancer diagnosis: A cohort study in the UK biobank." *British Journal of Cancer* 127:1534–1539.

High, K. A., and Roncarolo, M. G. 2019. "Gene therapy." *New England Journal of Medicine* 381:455–464.

Sanderson, E., Glymour, M. M., Holmes, M. V., Kang, H., Morrison, J., Munafò, M. R., et al. 2022. "Mendelian randomization." *Nature Reviews Methods Primers* 2:6. https://doi.org/10.1038/s43586-021-00092-5.

Tulchinsky, T. H. 2018. "John Snow, Cholera, the Broad Street Pump; Waterborne Diseases Then and Now." In *Case Studies in Public Health*, 77–99. Elsevier Press.

Wray, N. R., Lin, T., Austin, J., McGrath, J. J., Hickie, I. B., Murray, G. K., and Visscher, P. M. 2021. "From basic science to clinical application of polygenic risk scores: A primer." *JAMA Psychiatry* 78:101–109.

INDEX